Examples of Commutative Rings

Harry C. Hutchins

Polygonal Publishing House
80 Passaic Avenue
Passaic NJ 07055 USA

Calligraphy on pages 21 and 22 by Rita Danzig

Library of Congress Cataloging in Publication Data
Hutchins, Harry C.
    Examples of commutative rings.

    Bibliography: p.
    Includes indexes.
    1. Commutative rings. I. Title.
QA251.3.H87   512'.4   81-13780
ISBN 0-936428-05-8   AACR2

*Manufactured in the United States of America
by Braun-Brumfield*

Contents

Preface   v
Notational Conventions   vii

Preface

For many years, in algebra as in other branches of mathematics, there has existed a trend toward abstraction. Likewise, there is a tendency to prefer polished elegant accounts of structure theory. Unfortunately, such accounts, however lucid and beautiful they may be, sometimes conceal the roots of the subject, and they are frequently poor in explicit examples. What is gained in coherence and readability is often lost in concreteness.

However, it is difficult to imagine anyone reading or listening to such an account without becoming curious:

"Are there non-Noetherian Krull domains?"

"How hard is it to build a 4-dimensional valuation domain?"

"What is the completion of $\mathbf{Z}$ in the 10-adic topology?"

A serious student's study of theory must be accompanied by at least some acquaintance with examples; one does not spend a lifetime studying the hypothetical anatomy of curiously absent unicorns. In addition, research is frequently motivated by examples. If all known examples of A-rings are B-rings, this suggests a conjecture; and conversely, attempts to prove the conjecture may result in a counterexample.

One hypothetical user of this book might be a student attempting to learn commutative ring theory from some standard textbook, such as [Kaplansky 1974b] or [Matsumura 1970]. I hope these examples may help to illustrate theory and distinguish between concepts. Others, perhaps, may enjoy browsing; the zoo of commutative rings includes a number of strange beasts.

In Part I, I have presented a large body of information about commutative rings and their ideals. Although I have omitted the proofs, I have tried to provide references to proofs elsewhere.

As for the examples, I have attempted to maintain a uniformity of presentation. In many cases, not all of the statements made are proven; in some cases, the proofs would be trivial or repetitive, and in other cases, they would add too many pages. I have tried to include explicit examples, rather than theorems which establish the existence of examples (using, say, transfinite induction and someone's proven-to-exist-but-never-constructed example of a peculiar abelian group). I have not attempted to draw a distinction between examples and counterexamples, as the difference is mostly one of attitude; one man's pathology is another man's normalcy.

All of the rings discussed herein are assumed to be commutative, and (save in Chapter 5) all are assumed to contain 1.

As with any work of this sort, the choice of topics has been personal and idiosyncratic. I am making no claim to encyclopedic coverage of the subject, nor even to completeness in my coverage of the topics included. I am assuming that the reader is familiar with the concepts of ring, ideal, prime ideal, unit element, module, and a certain amount of field theory. In Chapter 3, Section F, some terminology of point-set topology is used; in Section G of the same chapter, tensor products are used. Chapter 4 is a brief account of homological dimension.

An earlier draft of this work constituted my Ph. D. dissertation at the University of Chicago in 1978. I would like to express my sincere gratitude to my advisor, Irving Kaplansky, whose advice and assistance have been extremely valuable. Several individuals, including Daniel Anderson, Graham Evans, Robert Gilmer, and Jack Ohm, read the earlier draft and offered helpful remarks. I would also like to thank the University of Chicago and Northern Illinois University for their support and aid, and I would like to thank Michael Weinstein.

Dekalb, Illinois   Harry C. Hutchins
June, 1981

Notational Conventions

Z is the ring of integers, occasionally regarded only as an additive group.

Q is the field of rational numbers, occasionally regarded only as an additive group.

R is the field of real numbers.

C is the field of complex numbers.

Indeterminates will be represented usually by uppercase letters near the end of the alphabet, such as U, V, W, X, Y, Z, or by such letters with subscripts, such as X_1, X_2, X_3, ...

$R[X, Y, \ldots, Z]$ is the ring of all polynomials in the indeterminates X, Y, ..., Z with coefficients in the ring R.

$R[[X]]$ is the ring of all formal power series in the indeterminate X with coefficients in the ring R.

$K(X, Y, \ldots, Z)$ is the field of all rational functions in the indeterminates X, $Y, \ldots, Z$ with coefficients in the field K.

Multiplication has usually been denoted by juxtaposition, i.e., the product of a and b is ab; but in a few instances it has been denoted using parentheses, $(a)(b)$, and occasionally with a dot, $a \cdot b$. The dot $\cdot$ has also been used to indicate a principal ideal; when several rings are present, (t) may be ambiguous, but $t \cdot R$ is not.

If we have a subring R of a ring T and an element a of T, $R[a]$ denotes the subring of T generated by R and a. Thus, for instance, $R[2X]$ is a subring of $R[X]$. If I is an ideal of T, then we will denote by $R + I$ the subring of T consisting of all elements $r + i$, where $r \in R$ and $i \in I$. Thus, e.g., $\mathbf{Z} + X \cdot \mathbf{Q}[[X]]$ is a subring of $\mathbf{Q}[[X]]$. (In a few cases, the larger ring T may not be explicitly specified.)

Part I
Definitions and Exposition

1

Preliminaries

A. Noetherian Rings

Historically, the study of commutative rings commenced with the consideration of rings of polynomials having coefficients in $\mathbf{Z}$ or in an arbitrary field. These rings, among others, have the following extremely useful property.

A ring R is said to be *Noetherian* if it satisfies either one of the (equivalent) conditions:

(i) Every ideal of R is finitely generated as an R-module.
(ii) The ideals of R satisfy the ascending chain condition, i.e., any chain $I_1 \subset I_2 \subset I_3 \subset \ldots$ must terminate.

The term honors E. Noether, whose papers helped found the theory of commutative rings.

Chains of prime ideals are of particular interest. A prime ideal P of a ring is said to be of *height n*, or $\mathrm{ht}(P) = n$, if there exists a chain $P_0 \subset P_1 \subset \ldots \subset P_n = P$ of prime ideals of R, and no longer such chain exists. In particular, a prime ideal is of height 0 if it contains no other prime ideals.

An ideal I is said to be *principal* if it is generated by a single element x, written $I = (x)$.

If R is a Noetherian ring, x is a nonunit in R, and P is a prime ideal minimal among all prime ideals containing (x), then $\mathrm{ht}(P) \leqslant 1$. Specifically, if x is *nilpotent* (that is, $x^m = 0$ for some $m > 0$), then $\mathrm{ht}(P) = 0$, and if x is not a zero-divisor, then $\mathrm{ht}(P) = 1$. This extremely powerful result, the *principal ideal theorem,* is due to W. Krull. There is a more elaborate version:[1]

If R is a Noetherian ring, I is a proper ideal of R generated by n elements, and P is a prime ideal of R containing I, with P/I of height k in R/I, then the height of P is at most $n + k$.

It follows from the principal ideal theorem that any prime ideal of a Noetherian ring is of some finite height.

In a non-Noetherian ring, even the simple version of the principal ideal theorem can fail spectacularly. Moreover, it is quite possible that a given prime ideal may not be of any finite height. If P is such a prime ideal, we say that $\mathrm{ht}(P) = \infty$.

A ring R is said to be of *Krull dimension n,* written $\dim(R) = n$ (possibly infinite), if $n = $ the supremum of $\mathrm{ht}(P)$ for all prime ideals P of R.

Even a Noetherian ring may be of infinite Krull dimension. Likewise non-Noetherian rings can be found in any dimension. Of course, the simple version of the principal ideal theorem holds trivially in any ring of dimension $\leqslant 1$.

An ideal I of a ring R is said to be *radical* if, whenever $x \in R$ and $x^k \in I$ for some $k \geqslant 1$, then $x \in I$. Clearly any prime ideal is radical. Any intersection of radical ideals is again a radical ideal. Since not every intersection of prime ideals is a prime ideal, it follows that not all radical ideals are prime.

In fact, any radical ideal is the intersection of all prime ideals containing it. If I is any ideal of a ring R, we define a related ideal $\mathrm{rad}(I)$, called the *radical* of I, by $\mathrm{rad}(I) = $ the intersection of all prime ideals containing I. An alternate (and equivalent) definition is that $x \in \mathrm{rad}(I)$ if and only if $x^k \in I$ for some $k \geqslant 1$.

The ideal $\mathrm{rad}((0))$ is the smallest radical ideal of the ring, and is sometimes called the *nil radical*; it consists of all the nilpotent elements of R.

If I is a radical ideal and we want to represent I as an intersection of prime ideals, we clearly can discard any prime ideals which are not minimal among all prime ideals containing I (that is, we can discard any prime ideals not *minimal over I*). It is convenient if there should happen to be only finitely many prime ideals minimal over I.

If R is Noetherian (or, in fact, if R merely satisfies the ascending chain condition on radical ideals, a weaker condition), then there are only finitely many

---

[1] For proofs of both versions of the principal ideal theorem, see [Kaplansky 1974b, Theorems 142 and 154]; also see [van der Waerden 1970, p. 147].

prime ideals minimal over I. The converse to this theorem is not quite true.[2]

An ideal I of a ring R is said to be *irreducible* if I is not the intersection of two properly larger ideals. Prime ideals are irreducible; indeed, a prime ideal is precisely an irreducible radical ideal.

An ideal I of a ring is said to be *primary* if, whenever $a \in R$, $b \in R$, and $ab \in I$, then either $a \in I$ or $b^k \in I$ for some $k \geqslant 1$. Irreducible ideals are primary.[3] If I is a primary ideal, rad(I) is a prime ideal. However, primary ideals may be reducible, and there are nonprimary ideals I for which rad(I) is a prime ideal.

If I is a primary ideal and $P = $ rad(I), we say that I is P-primary. If I and J are both P-primary ideals, then $I \cap J$ is also P-primary. If M is a maximal ideal of a ring R and I is any ideal of R with $M^k \subseteq I$ for some $k \geqslant 1$, then I is M-primary.

In a Noetherian ring, any ideal is an intersection of finitely many irreducible ideals. (This fact has little to do with rings; it is really a statement about lattices, specifically the lattice of ideals of a ring.) Such a representation, of an ideal as an intersection of finitely many irreducible ideals, may fail (in any of several ways) to be unique.

Thus, suppose that R is Noetherian and I is an ideal, so $I = L_1 \cap L_2 \cap \ldots \cap L_m$. Each ideal L_i is irreducible, and so each L_i is P_i-primary, for prime ideals $P_i = $ rad(L_i). Some of the L_i may be *redundant* in the sense that, e.g., L_1 is redundant if it contains $L_2 \cap \ldots \cap L_m$. Therefore, some of the ideals L_i may be discarded. Also, the prime ideals P_i may not be distinct. For each prime ideal P_i, intersect together all of the L_i which are P_i-primary. Thus we obtain (possibly reducible) primary ideals $K_1, \ldots, K_n$ such that

(i) K_i is Q_i-primary, for $Q_i = $ rad(K_i),
(ii)   the prime ideals Q_i are all distinct,
(iii) $I = K_1 \cap K_2 \cap \ldots \cap K_n$,
(iv)   the intersection in (iii) is irredundant.

The ideals K_i are called the *primary components* of I. If, for a given i, the prime ideal Q_i is minimal over I, then the component K_i is uniquely determined, and Q_i is called an *isolated prime* of I. Otherwise, Q_i is called an *embedded prime* of I, and the component K_i may not be unique. In any case, the number n and the prime ideals $Q_1, Q_2, \ldots, Q_n$ are uniquely determined by the ideal I.

---

[2] See [Kaplansky 1974b, Theorem 88] for a proof of the theorem. A partial converse can be found in [Kaplansky 1974b, p. 65, Exercise 25].

[3] See [van der Waerden 1970, p. 127 ff.] for a thorough discussion of primary decomposition of ideals.

It follows that in any Noetherian ring, there are only finitely many prime ideals minimal over a given ideal. In a non-Noetherian ring, this property, and almost any of the other familiar features of primary decomposition, may fail.

For the ideal I above, the prime ideals $Q_1, Q_2, \ldots, Q_n$ are called the *associated primes* of I, and the set $\{Q_1, \ldots, Q_n\}$ is denoted by Ass(I).

Suppose M is a finitely generated module over a Noetherian ring R. It is possible to discuss primary decomposition for submodules of M. If $x \in M$, we define the *annihilator* of x, denoted by Ann(x), to be the ideal of R consisting of all $t \in R$ with $tx = 0$. A prime ideal of R is an *associated prime* of M if $P =$ Ann(x) for some nonzero element x of M. The set of all associated primes of M is denoted by Ass(M).

If $0 \to N \to M \to M/N \to 0$ is a short exact sequence of R-modules, then the following is true: Ass(N) $\subseteq$ Ass(M) $\subseteq$ Ass(N) $\cup$ Ass(M/N). The containments may be proper.[4]

B. Domains

Let R be an integral domain with quotient field K. R is said to be *integrally closed* if, whenever an element x of K satisfies an equation $x^n + a_1 x^{n-1} + \ldots + a_n = 0$, with all coefficients a_i from R, x is necessarily an element of R. An equivalent definition is that R is integrally closed if R contains all elements x of K such that $R(x)$ is a finitely generated R-module.

While fields are obviously integrally closed, most domains are not. The intersection of any family of integrally closed domains, all contained in some large given field, is integrally closed.

For R and K as above, R is said to be *completely integrally closed* if, whenever elements a and x of K are such that $a \neq 0$ and $ax^k \in R$ for all $k \geqslant 1, x$ is necessarily an element of R. An equivalent definition is that R is completely integrally closed if it contains all elements x of K such that $R[x]$ is contained in a finitely generated R-module.

Completely integrally closed domains are integrally closed. The intersection of any family of completely integrally closed domains, all contained in some large given field, is completely integrally closed.

---

[4] See [Matsumura 1970, pp. 49–57] for a discussion of primary decomposition of submodules, including the connection between the two notions of associated prime. Also see [Kaplansky 1974b, pp. 54–58].

A ring R is Noetherian if and only if every submodule of any finitely generated R-module is finitely generated; thus an integrally closed Noetherian domain is completely integrally closed. However, not all completely integrally closed domains are Noetherian.

An element a of a domain R is said to be *irreducible* if it cannot be written as a product $a = bc$ of nonunit elements b and c of R. The nonzero element a of R is said to be a *principal prime* if the ideal (a) is prime. Any principal prime element is irreducible, but not every irreducible element is a principal prime. In any domain with the ascending chain condition on principal ideals, there must be irreducible elements; but even in a Noetherian domain, it is possible that there might not be any principal prime ideals.

A domain R is said to be *factorial* (or a *unique factorization domain*) if R satisfies either of the (equivalent) conditions:

(i)   Any nonzero element of R can be written as a product of principal primes and units.

(ii)   Any nonzero prime ideal of R contains a principal prime ideal.[5]

In any factorial domain, every prime ideal of height 1 is principal. Since there are domains which are not fields, but which nonetheless have no prime ideals of height 1, this condition is not equivalent. However, a Noetherian domain is factorial if and only if every prime ideal of height 1 is principal.

Any factorial domain is completely integrally closed and satisfies the simple version of the principal ideal theorem. However, a domain may be Noetherian and integrally closed and still not be factorial. A domain may be factorial and still fail to be Noetherian.

Let R be an integral domain with quotient field K. An R-submodule of K is called a *fractional ideal* of R. Oridinary ideals are clearly fractional ideals; these are sometimes referred to as *integral* fractional ideals. (Occasionally the term "fractional ideal" is used with a more restrictive definition; for instance, one might say that an R-submodule I of K is a fractional ideal if it is finitely generated, or if there is a nonzero $t \in R$ with $tI \subseteq R$.)

If I is a fractional ideal of R, we define $I^{-1} = \{x \mid xI \subseteq R\}$. If I and J are fractional ideals of R, we define their product $IJ = \{a_1 b_1 + a_2 b_2 + \ldots + a_m b_m$, with $a_i \in I$ and $b_j \in J$ for all i and $j\}$. A fractional ideal I is said to be *invertible* if $II^{-1} = R$. Notice that for any fractional ideal I we have $II^{-1} \subseteq R$.

If I is a fractional ideal generated as an R-module by one element (a *cyclic*

---

[5] One elegant proof that these conditions are equivalent can be found in [Kaplansky 1974b, Theorem 5]. There are also many other conditions equivalent to factoriality.

R-module), I is called a *principal* fractional ideal. We have $(x)^{-1} = (x^{-1})$, so any principal fractional ideal is invertible. In general, invertible fractional ideals are not necessarily principal. For each $n \geqslant 1$, there is a domain in which some invertible ideal may require n generators. If R is a factorial domain, however, invertible fractional ideals are principal.[6] This useful property also holds in some other classes of domains.

In any case, an invertible fractional ideal is always finitely generated as an R-module. The notion of "invertible fractional ideal" can be thought of as a generalization of that of "principal ideal."

C. Quasi-local Rings

A ring R is said to be *quasi-local* if it has exactly one maximal ideal; and is said to be *local* if it is Noetherian and quasi-local. By convention, one does not usually consider a field to be a quasi-local ring.

Quasi-local and local rings have useful properties, among which is that in general any ring can be made quasi-local, by a procedure known as *localization.*

Let R be a ring and let S be a multiplicatively closed subset of R. Let us assume that $0 \notin S$. We may also assume that S contains all of the units of R, and that S is *saturated,* i.e., S contains all divisors of elements of S. (The first assumption is necessary; the other two are harmless.)

Now we can construct a new ring, R_S, in which all of the elements of S have become units. R_S consists of equivalence classes of formal quotients a/s, where $a \in R$ and $s \in S$, and $a/s \sim b/t$ if and only if $u(at - bs) = 0$ for some element u of S. R_S becomes a ring once we define multiplication and addition (on typical representatives of equivalence classes) by $a/s \cdot b/t = ab/st$ and $a/s + b/t = (at + bs)/st$. There is a homomorphism $R \to R_S$, defined by sending each $x \in R$ to $x/1$ in R_S. This map may not be one-to-one and is surjective only in trivial cases.

Any proper ideal I of R which has an empty intersection with S "survives" in R_S, that is, there is an inclusion-preserving correspondence between such ideals of R and all of the proper ideals of R_S, where I corresponds to the ideal I_S of R_S generated by the image of I.

The ring R_S is called the *localization of R at S.* If R is a domain, R_S is actually a subring of the quotient field of R, and the inclusion of R into R_S is obvious-

---

[6] See [Kaplansky 1974b, p. 42, Exercise 15]. In fact, a stronger theorem is that in a pseudo-Bézout domain, invertible fractional ideals are principal.

ly one-to-one. In any ring R, there are several standard kinds of multiplicatively closed sets S:

(i) $S = \{x, x^2, x^3, \ldots\}$, for some nonnilpotent element x of R. If x is a non-zero-divisor, R_S is in this case sometimes denoted by $R[x^{-1}]$.

(ii) S = the set-theoretic complement of a prime ideal P of R. In this case, R_S is often denoted by R_P and is called R *localized at P.*

(iii) S = the intersection of $S_1, S_2, \ldots, S_m$, where each S_i is the complement of a prime ideal P_i.

If P is a prime ideal in R, the correspondence of ideals described above shows that R_P is a quasi-local ring, having as its maximal ideal P_P. If R is Noetherian, then R_P is a local ring.

If $P_1, \ldots, P_m$ are prime ideals of R and S is the intersection of their complements, then R_S has at most m maximal ideals, namely $P_{1S}, \ldots, P_{mS}$. (There might be fewer if, say, $P_1 \subseteq P_2$.) This fact follows from the following lemma:

If $P_1, \ldots, P_m$ are prime ideals in R and I is an ideal of R such that $I \subseteq P_1 \cup P_2 \cup \ldots \cup P_m$, then in fact $I \subseteq P_i$ for some i.[7] This lemma can be strengthened somewhat, but not in the direction of replacing "finitely many prime ideals" with "infinitely many prime ideals."

In any ring R, a fourth kind of multiplicatively closed set is

(iv) S = the set of all non-zero-divisors in R. In this case, R_S is called the *total quotient ring of R.*

Some rings are already their own total quotient rings, i.e., every nonunit is a zero-divisor. Any ring of dimension 0 is its own total quotient ring,[8] but not conversely.

(Often in the literature the term "localization" is reserved for R_P, while R_S is referred to as a *quotient ring.*)

A ring having only finitely many maximal ideals is called a *semi-quasi-local* ring. If it is Noetherian as well, it is called a *semi-local* ring.

Let R be a domain with quotient field K. Then $K = R_{(0)}$, since (0) is a prime ideal. There may be many subrings of K which are not localizations of R. In general, a subring of K which contains R is called an *overring* of R.

---

[7] This is a standard theorem. See [Kaplansky 1974b, Theorem 81], or [Matsumura 1970, p. 2] for a proof. See also [Quartararo and Butts 1975] for more information.

[8] See [Kaplansky 1974b, Theorem 91] for a proof.

If R is a semi-quasi-local domain, all invertible fractional ideals of R are principal. In fact, an ideal I of a domain R is invertible if and only if I is finitely generated and I_M is principal for every maximal ideal M of R.[9]

If R is a semi-local ring, then R is of finite Krull dimension, namely the supremum of the heights of all maximal ideals.

Suppose R is a quasi-local ring with maximal ideal M and suppose A is a finitely generated R-module. If $MA = A$, then $A = (0)$. This result is called the *Nakayama lemma* (and various other names).[10] It is very useful; some versions of it hold even in the noncommutative case.

Suppose R is a 1-dimensional, integrally closed, local domain. Then R has precisely one nonzero prime ideal, M, which turns out to be principal.[11] Such a ring is called a *Noetherian valuation domain*: it is in many ways the very nicest possible 1-dimensional ring. There is a generalization of this "niceness" to higher dimensions.

Let R be a local ring with maximal ideal M. Then M/M^2 is a module over R/M. Actually, R/M is a field, so M/M^2 is a vector space over R/M, of some finite dimension d. If $a_1, \ldots, a_d$ are elements of R, then $M = (a_1, \ldots, a_d)$ if and only if the images of $a_1, \ldots, a_d$ span M/M^2 over R/M. (One way to prove this is by way of the Nakayama lemma.)

It follows from the fancier version of the principal ideal theorem that $\dim(R) \leqslant d$. If $d = \dim(R)$, we say that R is a *regular local ring*. A Noetherian valuation domain is simply a regular local ring of dimension 1.

Any regular local ring is a factorial domain,[12] and thereby is integrally closed. By no means are all local domains, or even all local factorial domains, regular. Various attempts have been made to define regularity in the non-Noetherian quasi-local case.[13]

---

[9] Part of this theorem is trivial; see [Kaplansky 1974b, Theorem 62] for a proof.

[10] Proofs of the Nakayama lemma can be found in [Kaplansky 1974b, Theorem 78], or in [Matsumura 1970, p. 11], or in [Nagata 1962, p. 12]. See [Nagata 1962, p. 212] for a discussion of the lemma's history.

[11] See [Kaplansky 1974b, Theorem 95] for a proof.

[12] This is not easy to prove, although there are several weaker statements which are not as difficult: the 1-dimensional case, or the proof that a regular local ring is a domain. For a proof, see [Kaplansky 1974b, Theorem 184], or [Matsumura 1970, p. 142], or [Nagata 1962, p. 99].

[13] See item #1 in the Further References section.

Let R be any ring and let A be any R-module. An ordered sequence $x_1, x_2, \ldots, x_m$ of elements of R is said to be an *R-sequence on A* if

(i) $(x_1, \ldots, x_m)A \neq A$, and
(ii) x_1 is not a zero-divisor on A and for each $i \geq 2$, x_i is not a zero-divisor on $A/(x_1, \ldots, x_{i-1})A$.

The order of the elements is quite significant; in general, a permutation of an R-sequence is not an R-sequence. If, however, R is local and A is finitely generated, then any permutation of an R-sequence on A is also an R-sequence.[14]

If R is Noetherian, any R-sequence on A can be extended to an R-sequence of maximal length (still finite). If, in addition, I is an ideal of R, A is finitely generated, and $IA \neq A$, then any two R-sequences on A contained in I, which are maximal, are necessarily of the same length, which we shall denote by $Gr(I, A)$ and call the *grade of I on A*.

If R is a local ring, with maximal ideal M, we denote $Gr(M, R)$ by $Gr(R)$ and call this the *grade of R*. It is true in general that $\dim(R) \geq Gr(R)$. If $\dim(R) = Gr(R)$, we say that R is a *Cohen–Macaulay* local ring.

Any regular local ring is Cohen–Macaulay, but the converse is quite false. A local ring R which is Cohen–Macaulay may not be a domain; if it is a domain, it may not be integrally closed; if it is an integrally closed domain, it may not be factorial; and even if it is a factorial domain, it may not be regular. On the other hand, there are local factorial domains which are not Cohen–Macaulay.

If R is any Noetherian ring, we say that R is *regular* if R_M is a regular local ring for every maximal ideal M. Likewise, R is *Cohen–Macaulay* if R_M is a Cohen–Macaulay local ring for every maximal ideal M.

If R is a regular ring and t is a nonunit in R, then $R/(t)$ is regular if and only if t is not in the square of any maximal ideal. Likewise, if R is Cohen–Macaulay, then R/I is Cohen–Macaulay if I is generated by an R-sequence. These facts are useful in showing that various specific rings either are or are not regular or Cohen–Macaulay.

If R is a Noetherian ring, then for an ideal I, $I^{-1} = R$ if and only if $Gr(I, R) \geq 2$.

Any regular ring is Cohen–Macaulay. A regular ring may not be a domain, but at least it will be a direct sum of regular domains.[15] Regular domains need not be factorial, but they are integrally closed.

---

[14] For a discussion on R-sequences and grade, see [Kaplansky 1974b, p. 84 ff.].
[15] This is a special case of [Kaplansky 1974b, Theorem 168].

Any integrally closed Noetherian domain of Krull dimension 2 or less is Cohen–Macaulay. This cannot be improved, even to dimension 3.

It is useful here to notice that the Cohen–Macaulay property is defined only for Noetherian rings. If R is not Noetherian, the concept of grade becomes very intractable. There may be maximal R-sequences on a module which are of different lengths, or even of infinite length.[16]

Let R be any ring, and let P and Q be prime ideals of R with P contained in Q. A chain of prime ideals of R, all lying between P and Q is said to be *saturated* if no additional prime ideals can be inserted to make a longer chain. The ring R is said to be *catenary* if, for any two such prime ideals P and Q, any two saturated chains of prime ideals starting with P and ascending to Q are of the same finite length.

Any ring of dimension 2 or less is catenary, and any Cohen–Macaulay ring is catenary.[17] It is fairly easy to construct noncatenary rings with zero-divisors, or even integrally closed quasi-local domains which are not catenary. At much greater effort, Noetherian domains can be found which are not catenary; but the problem of constructing a noncatenary, integrally closed, Noetherian domain has only recently been solved.[18]

A ring R is said to be *universally catenary* if every polynomial ring $R[X_1, \ldots, X_m]$ in a finite number of indeterminates is catenary. If R is universally catenary, it is catenary, but the converse is false. There are examples of 1-dimensional rings R (which are trivially catenary) for which $R[X]$ is not catenary.

---

[16] See [Hochster 1974] for information on the notion of grade for non-Noetherian rings.

[17] "Catenary" is but one of several available chain conditions on saturated chains of prime ideals. See [Kaplansky 1974b, Theorem 138]; also see [Nagata 1962, p. 122 ff.].

[18] See item #2 in the Further References section.

2

Domains

A. Valuation Domains

Let R be an integral domain with quotient field K. R is said to be a *valuation domain* if it satisfies either of the (equivalent) conditions:

(i)   For any two elements x and y of R, either x divides y or y divides x.

(ii)   For any element x of K, either $x \in R$ or $x^{-1} \in R$.

Valuation domains are integrally closed.[1] A valuation domain is completely integrally closed if and only if it is 1-dimensional. A Noetherian valuation domain is 1-dimensional; but not all 1-dimensional valuation domains are Noetherian. Of course, any valuation domain is quasi-local; the maximal ideal consists of all nonunits, as in any quasi-local ring.

The ideals of a valuation domain R (in particular, the prime ideals) are linearly ordered by inclusion. Any finitely generated ideal of R (in particular, any

---

[1] For a proof that pseudo-Bézout domains are integrally closed, which is a stronger statement, see [Kaplansky 1974b, Theorem 50]. For material on valuations, see [Gilmer 1972a, Chapter III] or [Bourbaki 1972, Chapter VI].

invertible ideal) is principal. Radical ideals of R are prime ideals. The principal ideal theorem fails spectacularly in any valuation domain of dimension at least 2; if P is a nonzero nonmaximal prime ideal and $x \notin P$, then any prime ideal minimal over (x) must have height at least 2 (assuming x is not a unit).

The name "valuation domain" comes from the fact that classically valuation domains arise from the study of *valuations* on fields. If K is a field, a *valuation* on K is a map $v:K \to G \cup \{\infty\}$, where G is a linearly ordered abelian group, written additively, and ∞ is defined to be greater than any element of G, such that for all elements a and b of K,

(i) $v(ab) = v(a) + v(b)$,
(ii) $v(a) = \infty$ if and only if $a = 0$, and
(iii) $v(a + b) \geqslant$ the lesser of $v(a)$ and $v(b)$.

The group G is called the *value group* of the valuation v. If $R = \{x \mid x \in K$ and $v(x) \geqslant 0\}$, R is a valuation domain, sometimes denoted R_v. Its maximal ideal is precisely $M = \{x \mid x \in K$ and $v(x) > 0\}$.

There is an extensive theory of valuations, largely due to W. Krull. It is quite possible to obtain the same valuation domain by way of several distinct valuations.

Valuation domains exist in abundance. In particular, suppose R is a domain contained in a field K, and M is a maximal ideal of R. Then there is a valuation domain V with quotient field K and maximal ideal N such that $N \cap R = M$.[2]

If R is a Noetherian valuation domain, its value group is isomorphic to $\mathbf{Z}$. If R is 1-dimensional, then it is Noetherian if and only if its value group contains a least positive element, or if and only if the maximal ideal M of R has $M \neq M^2$. R is 1-dimensional if and only if its value group is isomorphic, under an order-preserving isomorphism, to a subgroup of the linearly ordered group of real numbers.

(A Noetherian valuation domain is often called a "discrete" valuation domain. Another usage of "discrete" is that a valuation domain R is *discrete* if, for any prime ideals P and Q of R with $P \subseteq Q$ and Q/P of height 1 in R/P, $(R/P)_{Q/P}$ is Noetherian. I will continue using "Noetherian," which is unmistakable.)

Just as the notion of "regular local ring" was globalized to "regular ring," so the notion of "valuation domain" has been globalized. A domain R is said to be a *Prüfer* domain if it satisfies either of the (equivalent) conditions:

---

[2] See [Kaplansky 1974b, Theorems 55 and 56, and p. 44, Exercise 34] for a proof.

(i)   Any finitely generated nonzero ideal of R is invertible.

(ii) R_M is a valuation domain for every maximal ideal M of R.

There are dozens, perhaps hundreds, of conditions equivalent to "Prüfer": conditions on ideals, on prime ideals, on modules, etc.[3]

Prüfer domains are integrally closed and catenary, and 1-dimensional Prüfer domains are completely integrally closed. Any overring of a Prüfer domain is Prüfer.

A Noetherian Prüfer domain is called a *Dedekind* domain. There are again many equivalent conditions: A domain R is a Dedekind domain if and only if

(i) R is Noetherian, integrally closed, and 1-dimensional, or

(ii)   every nonzero ideal of R is invertible, or

(iii)   every (proper) ideal of R is a product of prime ideals, or

(iv) R is 1-dimensional and regular.

A domain R is said to be *Bézout* if every finitely generated ideal of R is principal. Any valuation domain is Bézout. Since principal ideals are invertible, any Bézout domain is Prüfer. There are Bézout domains which are not valuation domains, and there are Prüfer domains which are not Bézout. A semi-quasi-local Prüfer domain is Bézout, but this sufficient condition is not necessary.

A Noetherian Bézout domain is called a *principal ideal domain.* Here are several equivalent definitions:

(i)   Each ideal of R is principal and R is a domain.

(ii) R is 1-dimensional and factorial.

(iii) R is Prüfer and factorial.

(iv)   Each prime ideal of R is principal and R is a domain.

Clearly, any principal ideal domain is a Dedekind domain. Both kinds of domains arise naturally and frequently in algebraic number theory. It is of considerable interest to know which Dedekind domains are principal ideal domains.

In any Dedekind domain, any ideal I can be generated by two elements, i.e., $I = (a, b)$ for some a and b. The parallel question for Prüfer domains—whether finitely generated ideals require no more than two generators[4]—has only recently been settled, in the negative.[5]

---

[3] See [Gilmer 1972a, Chapter IV] for much material on Prüfer domains.

[4] This is called the *2-generator* property.

[5] See items #3 and 4 in the Further References section.

Suppose R is an integral domain and x and y are nonzero elements of R. Then an element z is said to be the *greatest common divisor* of x and y, denoted $z =$ g.c.d.(x, y), if z divides both x and y, and for every element t which divides both x and y, t divides z. Of course, greatest common divisors are only unique up to multiplication by units of R.

If g.c.d.(x, y) exists for every x and y in R, R is said to be *pseudo-Bézout*.[6] Bézout domains and factorial domains are pseudo-Bézout. There are pseudo-Bézout domains which are neither factorial nor Bézout, in dimensions greater than 1; but a 1-dimensional pseudo-Bézout domain is actually Bézout.[7]

Pseudo-Bézout domains are integrally closed. If R is Noetherian and pseudo-Bézout, it is factorial; indeed, R is factorial if and only if it is pseudo-Bézout and satisfies the ascending chain condition on principal ideals. If R is pseudo-Bézout, its invertible ideals are principal. (If g.c.d.(x, y) exists, then $((x, y)^{-1})^{-1} =$ (g.c.d.(x, y))).) If R is pseudo-Bézout, irreducible elements of R are principal primes, but there are domains which are not pseudo-Bézout but in which all irreducible elements are principal primes.

One can define the *least common multiple* of x and y, or l.c.m.(x, y). In a domain (with 1), the existence of g.c.d.(x, y) for all x and y is equivalent to the existence of l.c.m.(x, y) for all x and y, and their product is xy. There are domains in which for some specific elements x and y, g.c.d.(x, y) exists, but l.c.m.(x, y) does not exist.

B. The Group of Divisibility

Suppose R is a domain with quotient field K. Let R^* and K^* be the multiplicative groups of units of R and K respectively. The groups R^* and K^* are abelian, and R^* is a subgroup of K^*. The *group of divisibility of R* is the factor group K^*/R^*, denoted by $G(R)$.

Notice that the value group of a valuation domain R is isomorphic to $G(R)$.

For any domain R, there is a natural partial order on $G(R)$, defined, for nonzero elements a and b of R, by $a^* \leq b^*$ if and only if a divides b in R, where a^* and b^* are the images of a and b, respectively, in $G(R)$. This order is then ex-

---

[6] See [Kaplansky 1974b, pp. 32–33] or [Gilmer 1972a, p. 75 ff.] for material on g.c.d.s.

[7] See [Sheldon 1974]. This is a nice little result. It actually suffices to prove that a 1-dimensional, quasi-local, pseudo-Bézout domain is a valuation domain.

tended to all of $G(R)$. Since a and b each divide ab, always, this is actually a directed partial order. (A partial order is *directed* if, for any two elements u and v, there is an element w with $u, v \leqslant w$.) It is not necessarily a lattice, however. (A partially ordered group is *lattice-ordered* if each finite subset has a least upper bound.)

Suppose R is integrally closed. Then $G(R)$ is a torsion-free group. (An abelian group is *torsion-free* if every element except the identity is of infinite order.) Actually, the precise condition needed is not integral closure; what is needed is *root-closure;* a domain R with quotient field K is *root-closed* if for any $x \in K$ such that $x^n \in R$ for some $n > 0$, then $x \in R$. This condition is weaker than integral closure.

Many of the properties of the domain R are related to the properties of $G(R)$. R is a valuation domain if and only if $G(R)$ is linearly ordered. R is pseudo-Bézout if and only if $G(R)$ is lattice-ordered. R has the ascending chain condition on principal ideals if and only if $G(R)$ has the descending chain condition on elements; that is, if and only if any chain $a_1 \geqslant a_2 \geqslant a_3 \geqslant \ldots$ must terminate. Putting together two of these statements, it can be shown that R is factorial if and only if $G(R)$ is a sum of copies of $\mathbf{Z}$. If T is any overring of R, then $G(T)$ is a homomorphic image of $G(R)$.[8]

Suppose G_0 is any lattice-ordered abelian group. Does a domain R exist having $G(R) \cong G_0$? The answer is yes, and one construction for such a domain is as follows:

Let K be any field, and let $T = K[t^g \text{ for all } g \in G_0]$, where $t^g \cdot t^h = t^{g+h}$. Thus T is the *group ring* of G_0 over K. A typical element of T is of the form $x = a_1 t^{g_1} + \ldots + a_m t^{g_m}$, where $a_1, \ldots a_m \in K$ and $g_1, \ldots, g_m \in G_0$. Let S be the subring of T consisting of all such elements x for which $\inf(g_1, \ldots, g_m) \geqslant 0$. Let $R = S_U$, where U is the multiplicatively closed set consisting of all elements of S having nonzero constant terms. (Notice that $t^0 = 1$.)

This domain in Bézout and has $G(R) \cong G_0$. The construction and its properties are due variously to Krull, Kaplansky, Jaffard, and Ohm.

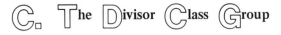 **C. The Divisor Class Group**

Let R be a domain with quotient field K. A fractional ideal I of R is said to

---

[8] See [Gilmer 1972a, Chapter III] for a great deal of material on $G(R)$; also see [Mott 1973] and [Mott 1974].

be *divisorial* if $(I^{-1})^{-1} = I$. Equivalently, a fractional ideal I is divisorial if and only if I is the intersection of all principal fractional ideals containing I.

Clearly, principal ideals and invertible ideals are divisorial; but in general, divisorial ideals are not invertible, and the intersection of even as few as two principal ideals, while it must be divisorial, does not have to be principal, invertible, or even finitely generated.

Let $D(R)$ be the set of all divisorial ideals of R. We define the product of two divisorial ideals I and J to be $I \cdot J = ((IJ)^{-1})^{-1}$. Then $D(R)$ is an abelian monoid with R as its identity.

R is completely integrally closed if and only if $D(R)$ is a group. In any case, $D(R)$ contains a subgroup $\text{Prin}(R)$, consisting of all principal fractional ideals. Suppose $D(R)$ is a group. Then the factor group $D(R)/\text{Prin}(R)$ is called the *divisor class group* of R and is denoted by $\text{Cl}(R)$.[9]

There is another way to obtain $D(R)$ which explains the name "divisor class group." For any fractional ideal I of R, define the *divisor* of I to be $\text{div}(I) =$ the intersection of all principal fractional ideals containing I. (Or equivalently, $\text{div}(I) = (I^{-1})^{-1}$.) Thus $\text{div}(I)$ is the smallest divisorial ideal containing I. Declare two fractional ideals I and J to be equivalent if $\text{div}(I) = \text{div}(J)$. $D(R)$ can be thought of as the set of all such equivalence classes, each represented by a divisorial ideal.

The divisor class group is of most interest in studying one particular class of domains. Let R be a domain with quotient field K. R is said to be a *Krull* domain if there exists a family of Noetherian valuation domains, $\{ V_i \mid i \in S \}$, each containing R and contained in K, such that R is the intersection of the V_i and each nonzero element of R is a unit in all but finitely many of the V_i.

If R is a Krull domain and the V_i are as above, then each V_i is R_P for some prime ideal P of R of height 1, and conversely, if P is a prime ideal of R of height 1, then R_P is a Noetherian valuation domain, one of the V_i.

An equivalent definition is that R is a Krull domain if and only if it is completely integrally closed and it satisfies the ascending chain condition on divisorial ideals. If R is a Krull domain, $D(R)$ is generated by the divisor classes of the prime ideals of height 1.

There are various generalizations available of the notion of Krull domain. For instance, the valuation domains V_i might be required to be merely 1-dimensional, not Noetherian. There is a limit to this sort of generalization, how-

---

[9] For material on divisors, Krull domains, and $\text{Cl}(R)$, see [Bourbaki 1972, Chapter VII] or [Fossum 1973].

ever, since any integrally closed domain is the intersection of all valuation domains containing it—and there are integrally closed domains which are very far from being Krull.

Krull domains exist in abundance. Any Noetherian integrally closed domain is Krull, and so is any factorial domain.

A Dedekind domain is simply a 1-dimensional Krull domain. A domain R is factorial if and only if R is a Krull domain and $Cl(R) = (0)$. Thus the divisor class group can be thought of as a way to measure how far a domain is from being factorial. For a Krull domain to be factorial, every divisorial ideal must be principal; it is not sufficient for every invertible ideal to be principal.

One fact, frequently useful in working with divisors, is the following: If I is an ideal of a domain R and $Gr(I, R) \geqslant 2$, then $I^{-1} = R$. If R is Noetherian, then the converse is also true.[10]

Just as with groups of divisibility, it is natural to ask which abelian groups can occur as divisor class groups. This important question was answered by L. Claborn in 1966: Let G_0 be any abelian group. Then there exists a Dedekind domain R having $Cl(R) \cong G_0$.[11]

The original proof of this theorem is rather involved and is not nearly as explicit as the previous construction for groups of divisibility. It is in some ways the best such theorem one could want. (It is also true that G_0 could be realized as $Cl(R)$ for a Krull domain R of arbitrary dimension.)

If R is a Prüfer domain, R may fail to be completely integrally closed, so $D(R)$ may fail to be a group. Let $I(R)$ be the set of all invertible ideals of R (which is to say, all the finitely generated ideals). Then $I(R)$ is a subgroup of $D(R)$, containing $Prin(R)$, and the quotient group $I(R)/Prin(R)$ is called the *class group* of R. (It is also true that any abelian group can be realized as the class group of a suitable Prüfer domain.)

D. Miscellaneous Properties

Let R be a domain with quotient field K. An element J of $D(R)$ is called a

---

[10] See [Kaplansky 1974b, p. 102, Exercises 1–2].

[11] See [Claborn 1966]; or see [Fossum 1973, Chapter III] for a lucid account of Claborn's theorem. In recent years, several alternate constructions have been discovered; for instance see [Eakin and Heinzer 1973] and Further References, item #9.

finitely generated divisor if $J = \text{div}(I)$ for some finitely generated fractional ideal I.

Suppose that R is not completely integrally closed. Then $D(R)$ is not a group, but merely a monoid. If every finitely generated divisor J is *regular* in $D(R)$, i.e., if $J \cdot L_1 = J \cdot L_2$ implies $L_1 = L_2$ for all L_1 and L_2 in $D(R)$, then R is said to be *regularly integrally closed*.[12]

Since invertible elements of a monoid are regular, it follows that a completely integrally closed domain is regularly integrally closed. If R is regularly integrally closed, then it is integrally closed.

If the set of all finitely generated divisors is actually a subgroup of the monoid $D(R)$, then R is said to be *pseudo-Prüfer*.[13] Any pseudo-Prüfer domain is regularly integrally closed. Any Krull domain or Prüfer domain or pseudo-Bézout domain is pseudo-Prüfer. None of these statements can be reversed; in fact, it is possible for a domain to be completely integrally closed but not pseudo-Prüfer.

Suppose R is a pseudo-Bézout domain. Then $G(R)$ is a lattice under its natural order. R is said to be *pseudo-principal* if it satisfies one of the (equivalent) conditions:

(i) $G(R)$ is a *complete* lattice, i.e., any nonempty bounded subset has a supremum.
(ii)   Any nonempty set of nonzero elements of R has a greatest common divisor.
(iii) $\text{Cl}(R)$ is a group, namely (0).
(iv)   Every divisorial fractional ideal of R is principal.

By definition, any pseudo-principal domain is pseudo-Bézout and completely integrally closed. Any factorial domain is pseudo-principal, but not conversely.

A domain R is said to be *almost Krull* if R_M is a Krull domain for each maximal ideal M of R. An almost Krull domain is completely integrally closed, but it need not be a Krull domain. (The difference is a rather delicate one.)[14]

---

[12] For material on regularly integrally closed, pseudo-Prüfer, and pseudo-principal domains, see [Bourbaki 1972, Chapter VII (exercises)].

[13] [Gilmer 1972a] refers to pseudo-Prüfer domains by the name "Prüfer v-multiplication rings."

[14] See [Gilmer 1972a, Chapter VI] for material on almost Krull, almost Dedekind and π-domains. Is every almost Krull domain pseudo-Prüfer?

Likewise, a domain R is said to be *almost Dedekind* if R_P is a Noetherian valuation domain for each nonzero prime ideal P of R.

If R is almost Dedekind, it is very easy to see that it is 1-dimensional, Prüfer, and almost Krull. If R is 1-dimensional and almost Krull, then it is almost Dedekind. For Noetherian domains, "almost Krull" and "almost Dedekind" become "Krull" and "Dedekind" respectively.

There are various generalizations of factorial domains. Here are two of them:

A ring R is said to be *locally factorial* if R_M is factorial for each maximal ideal M of R.

A domain R is said to be a *π-domain* if it satisfies one of the (equivalent) conditions:

(i) R is Krull and locally factorial.
(ii) R is Krull and each prime ideal of height 1 is invertible.
(iii) Each principal ideal of R is a product of prime ideals.

By definition, any π-domain is locally factorial. Any factorial domain or regular domain is a π-domain. There are domains which are locally factorial, but not Krull and thus not π-domains. A locally factorial domain is clearly almost Krull, but the converse is false.

One observation to be made with regard to the definitions above is that in a Krull domain an invertible prime ideal is necessarily of height 1.

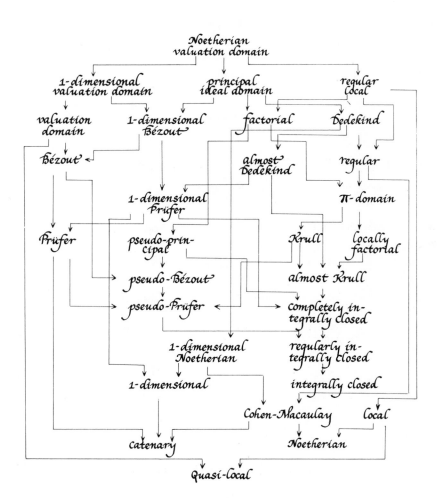

Properties of Domains

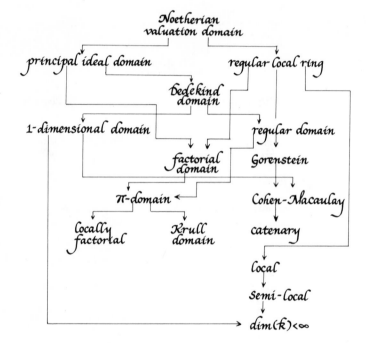

Properties of Noetherian Domains

3

Ring Extensions

A. Definitions

A large portion of commutative ring theory deals with situations of the following type: Suppose R and T are rings, related in some fashion (perhaps R is a subring of T, or T is a homomorphic image of R). If R has some property, does T also have it?

For most of this chapter, except where otherwise stated, we shall assume that R and T are rings and that R is a subring of T. In this case, the question above is referred to as the *ascent* question. The parallel question, namely, whether R must have some given property, given that T has it, is called the *descent* question.

Often we want to relate the prime ideals of R to the prime ideals of T. If Q is a prime ideal of T, then $Q \cap R$ is always a prime ideal of R. Can every prime ideal of R be obtained in this way from a prime ideal of T? There are several variations on this question.

We say that *lying-over* (abbreviated LO) holds for a pair of rings, R, T if for every prime ideal Q_0 of R, there is a prime ideal Q of T with $Q \cap R = Q_0$.

We say that *going-up* (abbreviated GU) holds for the pair R, T if whenever P_0 and Q_0 are prime ideals of R, $P_0 \subseteq Q_0$, and P is a prime ideal of T with $P \cap R = P_0$, then there exists a prime ideal Q of T with $P \subseteq Q$ and $Q \cap R = Q_0$.

It is interesting to notice that if P is a minimal prime ideal in R (i.e., minimal over rad$((0))$, or of height 0), then there is always a prime ideal in T lying over P (pick any ideal of T which is maximal with respect to excluding the complement of P in R; it will be prime). Because of this, LO holds for any pair of rings whenever GU holds.

We say that *going-down* (abbreviated GD) holds for the pair R, T if whenever P_0 and Q_0 are prime ideals of R, $P_0 \subseteq Q_0$, and Q is a prime ideal of T with $Q \cap R = Q_0$, then there exists a prime ideal P of T with $P \subseteq Q$ and $P \cap R = P_0$.

We say that *incomparability* (abbreviated INC) holds for the pair R, T if whenever P and Q are prime ideals of T with $P \cap R = Q \cap R$, then $P = Q$.

B. Localization

Here we consider the pair R, T where R is a ring, S is a multiplicatively closed subset of R, and $T = R_S$.

If R is Noetherian, integrally closed, factorial, regular, Cohen–Macaulay, Krull, Prüfer, Bézout, pseudo-Bézout, or a valuation domain, then R_S also has the same property.

In general, GD and INC hold for localizations, but LO (and therefore GU) fails. Since there is a one-to-one inclusion-preserving correspondence between the prime ideals of R_S and those prime ideals of R which have empty intersections with S, it is easy to see that $\dim(R_S) \leqslant \dim(R)$. If $R_S = R_P$ for some prime ideal P of R, we have $\dim(R_S) = \text{ht}(P)$.

Suppose R is a completely integrally closed domain, but not Krull. R_S may fail to be completely integrally closed.

If R is any domain, then R_S is a domain, an overring of R, and $G(R_S)$ is a homomorphic image of $G(R)$. On the other hand, not every overring of R is a localization (i.e., R need not satisfy the *QR property*), or even an intersection of localizations (i.e., R need not satisfy the *QQR property*), unless R has various properties, such as being Prüfer; specifically, Bézout implies QR, which implies Prüfer, which implies QQR.[1] If R is QQR and either finite-dimensional or integrally closed, then it is Prüfer. If R is Prüfer and has a torsion class group, then it is QR; but not all QR domains have torsion class groups. (An abelian group is *torsion* if it has no elements of infinite order.)

---

[1] See [Gilmer 1972a, pp. 334–340] and [Gilmer 1973].

There are relatively few descent results available for localizations. This should be obvious; if R is any domain whatsoever, $R_{(0)}$ is its quotient field and therefore has all sorts of nice properties.

Suppose R is a domain satisfying the ascending chain condition on principal ideals, and S is a multiplicatively closed subset of R, generated by a set of principal primes. Then if R_S is integrally closed, completely integrally closed, Krull, or factorial, R has the same property.[2]

Suppose R is semi-quasi-local and R_M is Noetherian for each maximal ideal M of R. Then R is Noetherian. Suppose R is a semi-local Krull domain and R_M is factorial for each maximal ideal M of R. Then R is factorial.

If R is not a domain, R_S may well be a domain, and different localizations may even have different characteristics.

ⓒ. ℙolynomials

Here we consider the pair R, T where R is a ring and T is the ring of all polynomials in some indeterminates. We shall be concerned primarily with the simplest case, $T = R[X]$.

R is Noetherian, integrally closed, completely integrally closed, Krull, pseudo-Bézout, factorial, Cohen–Macaulay, or regular, if and only if $R[X]$ also has each property respectively.[3]

In general, if P is a prime ideal of R, there are exactly two prime ideals of $R[X]$ which lie over P; namely $PR[X]$ and (P, X). Since $PR[X] \subseteq (P, X)$, INC fails for polynomial ring extensions. However, GD and LO always hold for the pair R, $R[X]$; and GU, except for the trivial case of $\dim(R) = 0$, usually fails.

If P is a prime ideal of R, then (as seen above) there is in $R[X]$ a chain of two prime ideals Q_1 and Q_2 such that $Q_1 \subseteq Q_2$ and $Q_1 \cap R = Q_2 \cap R = P$. However, no chain of three such prime ideals can exist in $R[X]$. From this, one can deduce that if $\dim(R) = n$, then $n + 1 \leqslant \dim(R[X]) \leqslant 2n + 1$. It is quite possible for $\dim(R[X])$ to attain the bounds in this inequality. In several nice cases, we have $\dim(R[X]) = n + 1$; specifically, if R is Noetherian or

---

[2] For a proof, consult [Kaplansky 1974b, Theorem 177] and [Gilmer 1972a, pp. 555–556, Exercise 11].

[3] The various parts of this statement are each more or less standard; proofs can be found scattered throughout [Bourbaki 1972], [Gilmer 1972a], and [Kaplansky 1974b].

0-dimensional, or if R is a Prüfer domain.

If R is any ring, we define the *dimension sequence* of R to be a_0, a_1, a_2, $\ldots$, where $a_0 = \dim(R)$ and $a_n = \dim(R[X_1, \ldots, X_n])$, for $n > 0$.

If R is Noetherian (or 0-dimensional or a Prüfer domain), the dimension sequence of R is simply a sequence of consecutive positive integers. If R is not Noetherian, the situation is much more chaotic.[4] We always have $a_0 + n \leqslant a_n \leqslant (a_0 + 1)(n + 1) - 1$, but there are other conditions as well. For instance, for any ring R, the differences $a_{n+1} - a_n$ are eventually constant (i.e., for $n > N$, these differences are all equal), but this eventual constant difference need not be 1.

If R is a Krull domain, then $Cl(R) \cong Cl(R[X])$. In $R[X]$, each divisorial ideal is div(P) for some prime ideal P of height 1.[5] This latter statement need not be true in an arbitrary Krull domain. (These two facts are both useful in one proof of Claborn's theorem.)

Suppose R is any ring. An element f of $R[X]$ can be written as $f = a_0 X^n + a_1 X^{n-1} + \ldots + a_n$. We can consider f as a function from R to R, where $f(c) = a_0 c^n + \ldots + a_n$. It is true, in general, that $X - c$ divides f if and only if $f(c) = 0$; but unless R is a domain, f may have more than n linear factors, or more than n roots, in $R[X]$.

Suppose R is an n-dimensional ring. Let $R_m = R[X_1, \ldots, X_m]$. It is true, in general, that $m + n \leqslant \dim(R_m) \leqslant (m + 1)(n + 1) - 1$, and there may be, in R_m, a chain of $m + 1$ prime ideals, all lying over the same prime ideal in R. However, it is quite possible that maximal ideals in R_m, or even in R_1, may be of various heights, even if R is Noetherian or local.

Rings of polynomials over a field are of especial interest. Let K be a field and let $R = K[X_1, X_2, \ldots, X_m]$. R is called an *affine ring*, and so is any homomorphic image of R (for any value of m). Affine rings are clearly Noetherian and catenary, and occur naturally and abundantly in algebraic geometry. In the ring R, all maximal ideals are of the same height, namely m. This makes it easy to find $\dim(T)$ for any affine domain T. Express T as $T = R/P$; then $\dim(T) = m - ht(P)$. What is often easier is to find the transcendence degree of the quotient field L of T over K. (Note that L must contain K.) Then we have $\dim(T) = $ tr. deg.(L/K).[6] (The *transcendence degree* of L over K is n, which is denoted

---

[4] See [Arnold and Gilmer 1974a] for a discussion of the possible dimension sequences for a ring.

[5] For proofs, see [Fossum 1973, p. 36 and p. 63].

[6] A somewhat stronger statement of this sort can be found in [Matsumura 1970, pp. 84–85].

tr. deg. $(L/K) = n$, if L is isomorphic to an algebraic extension of $K(Y_1, Y_2, \ldots, Y_n)$.)

Much research has been done on the question of which subrings of an affine ring are affine, or in particular, which subrings of a polynomial ring are polynomial rings.

It is possible to have two nonisomorphic rings R and T such that $R[X]$ and $T[X]$ are isomorphic.

D. Integral and almost Integral Extensions

Suppose T is a ring and R is a subring of T. An element t of T is said to be *integral* over R if it satisfies one of the (equivalent) conditions:

(i) t is the root of a monic polynomial with coefficients in R.
(ii) The subring $R[t]$ is a finitely generated R-submodule of T.

The ring T is said to be *integral* over R if every element of T is integral over R. The set of all elements t of T which are integral over R is a subring of T, called the *integral closure of R in T*.

Notice that any idempotent (an element t is *idempotent* if $t^2 = t$) or nilpotent element of T is integral over R. Thus, it is quite possible for a ring with zero-divisors to be integral over a domain. Of course, if T is a domain, so is R.

It is quite possible for R to be Noetherian, but not T, or for T to be Noetherian, but not R.

One of the most useful properties of integral extensions is that if T is integral over R, then LO, GU, and INC hold for the pair.[7] It follows from these that for any prime ideal P of R, there is a prime ideal Q of T with $Q \cap R = P$ and ht$(Q) =$ ht(P). To obtain such a Q, pick a chain of prime ideals $P_0 \subseteq P_1 \subseteq \ldots \subseteq P$, of length ht$(P)$, and use LO to find a prime ideal Q_0 of T lying over P_0. Then use GU to find a prime ideal Q_1 lying over P_1, and likewise $Q_2, \ldots, Q$ lying over $P_2, \ldots, P$ respectively. On the other hand, there may be several (or infinitely many) prime ideals Q in T lying over P. In any case, ht$(Q) \leqslant$ ht(P) and dim$(T) =$ dim(R).

Suppose T is an arbitrary integral extension of R. Aside from dimension, there are few if any properties of R which must be shared by T. As for descent

---

[7]See [Kaplansky 1974b, Theorem 44] or [Matsumura 1970, pp. 33–34].

questions, if T is a domain or quasi-local or semi-quasi-local, R must have the same property. Many other properties may fail; T may be (for instance) a regular factorial domain while R is neither integrally closed nor Noetherian.

If T is integral over R and is also a finitely generated R-algebra, then it is actually a finitely generated R-module. In this case, it is easy to see that if R is Noetherian, then so is T. It is a result of Eakin (and of Nagata, independently) that if T is Noetherian, so is R.[8] In addition, if P is a prime ideal of R, there are in T only finitely many prime ideals lying over P. Aside from these few facts, R and T still may have very different properties.

Suppose R is a domain and K is its quotient field. Let T be the integral closure of R in K. T is usually called the *integral closure* of R. T is never a localization of R unless $R = K$ or T. T is an integrally closed domain, so $T = R$ if and only if R is integrally closed to begin with. If R is Noetherian, T must be a Krull domain, but need not be Noetherian. If R is 1-dimensional and Noetherian, then T is actually a Dedekind domain. If R is 2-dimensional and Noetherian, then T must be Noetherian, but there may be a non-Noetherian ring contained in T and containing R. If R is Noetherian, regardless of dimension, and P is a prime ideal of R, then there are only finitely many prime ideals of T lying over P.

It is worth noticing that not every Krull domain is the integral closure of a Noetherian domain.

T is the intersection of all valuation domains containing R and contained in K. If R is not Noetherian, T may fail to be a Krull domain. If T is Cohen–Macaulay, R may not even be Noetherian, much less Cohen–Macaulay.

Let R be a domain with quotient field K. Let L be a field containing K. Let T be the integral closure of R in L. The precise relationship between R and T depends very much on the extension L of K.

If K is algebraically closed in L, then T is just the integral closure of R in K. For this reason, this situation is usually studied with the further assumption that R is integrally closed. (One can then separately study the relationship between a domain and its integral closure.) Let us assume that R is integrally closed.[9]

Suppose that L is an algebraic extension of K, possibly infinite-dimensional. If R is Noetherian, T may nonetheless fail to be Noetherian. In fact, even if R is a Noetherian valuation domain, T may fail to be Noetherian. Thus if R is a

---

[8] See [Eakin 1968] or [Kaplansky 1974b, p. 54, Exercise 15].
[9] Most of this material can be found in [Bourbaki 1972, Chapters V, VI, and VII (exercises)].

Krull domain, T may fail to be a Krull domain. On the other hand, if R is Prüfer, so is T. If R is Bézout, T must be Prüfer, but need not be Bézout; this can fail even in the 1-dimensional case, i.e., R may be a principal ideal domain while T is not Noetherian or factorial (or Bézout or pseudo-Bézout).

Conversely, if T is Krull and K_1 is any subfield of L, then $T \cap K_1$ is a Krull domain (specifically R is Krull). Thus, if T is a Dedekind domain, so is R. In general, however, T may be Noetherian or factorial while R is not.

In any case, T is completely integrally closed if and only if R is.

Next, let us assume that L is a finite-dimensional algebraic extension of K. This is the case that has received the most attention. If R is a Dedekind domain, then so is T. Even so, T may fail to be a finitely generated R-module. If R is a Krull domain of any dimension, then T is also a Krull domain. It is possible for either R or T to be a principal ideal domain while the other is not; thus in any dimension, either R or T may be factorial while the other is not. In any dimension, if R is Noetherian and L is a separable extension of K, then T is a finitely generated R-module and so is Noetherian. If L is not separable, T may not be a finitely generated R-module.[10]

The special case in which $R = \mathbf{Z}$ and $K = \mathbf{Q}$ is of great interest. L is a finite algebraic extension of $\mathbf{Q}$ and T is a Dedekind domain. It was the study of this situation which led to a great deal of classic algebraic number theory. Even if T is a simple extension of $\mathbf{Q}$, so $L = \mathbf{Q}[t]$, T need not be generated over $\mathbf{Z}$ by only one element—not an obvious fact. In this case, it is also true that $\mathrm{Cl}(T)$ is a finite group, which is emphatically not true of all Dedekind domains. Of course, $\mathrm{Cl}(\mathbf{Z})$ is trivial.

For any arbitrary field extension L of K, there is an intermediate field K_1 such that $K \subseteq K_1 \subseteq L$, L is algebraic over K_1, and K is algebraically closed in K_1.

There is an analogous theorem for affine rings. If $R = K[x_1, x_2, \ldots, x_n]$ is an affine ring, then there are elements $y_1, y_2, \ldots, y_m$ in R such that R is integral over $K[y_1, \ldots, y_m]$ and the elements y_i are algebraically independent over K.[11]

Suppose R is an integrally closed domain and T is a domain integral over R. Then GD holds for the pair R, T: that is to say, GU, GD, LO, and INC all hold

---

[10] Also see item #13 in the Further References section.

[11] This is often called the *normalization theorem,* and is due originally to E. Noether. See [Matsumura 1970, p. 91] for a proof.

for the pair R, T.[12] Thus, if P is a prime ideal of T, $ht(P) = ht(P \cap R)$, and given a chain of prime ideals in R, a chain of prime ideals in T lying precisely over it can be constructed starting from either end.

Integral extensions are of great interest in commutative ring theory. The related concept of an *almost integral extension* has not been studied as extensively. If R is a subring of T, an element t of T is said to be *almost integral* over R if t satisfies one of the (equivalent) conditions:

(i)   There is a non-zero-divisor $b \in R$ such that $bt^n \in R$ for all $n > 0$.

(ii)   The subring $R[t]$ is contained in a finitely generated R-submodule of T.

Notice that any element of R is almost integral over R, and any element of T which is integral over R is certainly almost integral over R.

Clearly, if R is Noetherian, the concepts of "integral" and "almost integral" coincide. On the other hand, it is quite possible to have $T = R_S$, a localization (very far from an integral extension), and have every element of T almost integral over R.

If every element of T is almost integral over R, T is said to be *almost integral* over R. The set of all elements of T which are almost integral over R is a subring of T, called the *complete integral closure* of R in T.

If R is a domain with quotient field K, let us denote the complete integral closure of R in K by R'. It is unfortunately true that R' may fail to be completely integrally closed. In fact, if $R^{(2)} = (R')'$ and in general $R^{(n+1)} = (R^{(n)})'$, then $R^{(n)}$ may fail to be completely integrally closed.[13]

Because almost integral extensions include both integral extensions and some localizations, very different phenomena, they have few useful properties. The prime ideals of T may have little to do with the prime ideals of R; GD, GU, LO, and INC may each fail. If $\dim(R)$ is finite, $\dim(T)$ may be infinite or any positive integer. If $\dim(R)$ is infinite, $\dim(T)$ may be infinite or any nonnegative integer.

If R is a valuation domain, then R' is completely integrally closed. Specifically, if R has a prime ideal Q with $ht(Q) = 1$, then $R' = R_Q$. Otherwise, R' is the quotient field of R.

---

[12] This fact is known as the *going-down theorem*. See [Matsumura 1970, pp. 31–34].

[13] See item #5 in the Further References section.

E. Power Series

Let R be a ring and let $T = R[[X]]$. We shall be concerned primarily with rings of power series in only one variable.

A typical element of T is of the form $b = a_0 + a_1 X + a_2 X^2 + \ldots$, where each a_i is an element of R. It is important to remember that this is a formal sum; there is no convergence involved. If the ring R is a topological ring (perhaps a field with a valuation), then a ring of convergent power series can be defined, but in general this is not $R[[X]]$.

R is Noetherian, Cohen–Macaulay, regular, Krull, or completely integrally closed if and only if $R[[X]]$ has each property respectively. If $R[[X]]$ is factorial, pseudo-Bézout, or integrally closed, then R also has each property respectively; for each of these, the converse fails. But R is a regular factorial domain if and only if $R[[X]]$ is also a regular factorial domain.

Suppose P is a prime ideal of R. There are always at least two prime ideals of $R[[X]]$ lying over P, just as in the polynomial case, namely (P, X) and $PR[[X]]$. There may be many others as well; in fact, there may be an infinite chain of prime ideals of $R[[X]]$, all lying over P. Thus INC fails for power series extensions. Just as with polynomial rings, GD and LO always hold for the pair R, T, and GU always fails unless $\dim(R) = 0$.

Let us consider $\dim(T)$. It is always true that $\dim(T) \geqslant \dim(R) + 1$. If R is Noetherian or 0-dimensional, then $\dim(T) = \dim(R) + 1$. In contrast to the case of polynomial rings, it is possible for $\dim(T)$ to be infinite even if R is Prüfer or a valuation domain or even a 1-dimensional valuation domain.[14]

It seems to be an open question whether there exists a finite-dimensional ring R for which $\dim(R[[X]])$ is finite but not $\dim(R) + 1$.

Suppose R is a domain with quotient field K. $K[[X]]$ is a Noetherian valuation domain with maximal ideal (X). Its quotient field is $K[[X]][X^{-1}]$, which is called the *field of Laurent series* in X over K. This field is denoted by $K((X))$.

The quotient field of $R[[X]]$ is usually much smaller than $K((X))$. (It obviously contains $K(X)$.) In fact, these are equal if and only if each countable family of nonzero ideals of R has a nonzero intersection. There is a valuation domain known which satisfies this rather strong condition.[15]

---

[14] See [Gilmer 1975] for information on $\dim(R[[X]])$.

[15] See [Gilmer 1967b].

In recent years, a considerable amount of work has been done toward characterizing the zero-divisors of $R[[X]]$, or the nilpotent elements of $R[[X]]$, or the automorphisms of $R[[X]]$ which leave fixed all of the elements of R.[16]

F. Completions

The subject of completions can be treated in very great generality, but we shall be concerned primarily with one case.[17]

Suppose R is a ring and I is a proper ideal of R such that $\cap I^n = (0)$. Then we can define a topology on R by defining $I, I^2, I^3, \ldots$ to be open neighborhoods of 0, and taking cosets $b + I^n$ to be open neighborhoods of b, for each $b \in R$. Using these sets as basic open sets, a topology is generated such that:

(i) R, with this topology, is Hausdorff.

(ii)   The operations of addition, multiplication, and negation are continuous.

Thus, R is a *topological ring.* This topology is called the *I-adic topology* on R. It is used most often when R is local and I is its maximal ideal, or when R is semi-local and I is its Jacobson radical. (The *Jacobson radical* of a ring is the intersection of all of its maximal ideals, and may be much larger than the nil radical.)

The set of all (equivalence classes of) Cauchy sequences of elements of R is denoted by $\overline{R}$ and bears a natural ring structure; $\overline{R}$ is called the *completion* of R. R is said to be *complete* in the I-adic topology if $R = \overline{R}$. It is sometimes necessary to consider the completions of R with respect to several different topologies; these usually yield nonisomorphic rings.

R is a subring of $\overline{R}$ in a natural way, by considering each $b \in R$ as a constant sequence. (This is one of several points that go awry if $\cap I^n = J \neq (0)$; if $J \neq (0)$, and one follows through the usual construction, what one obtains is the completion of R/J in the I/J-adic topology.)

---

[16] As with polynomial rings, it is possible to have two nonisomorphic rings R and S for which $R[[X]]$ and $S[[X]]$ are isomorphic. (Curiously, in this case, the Jacobson radicals of R and S must be nonzero.)

[17] Material on completions may be found in [Dieudonne 1967] and in [Matsumura 1970]. Material on point-set topology may be found in [Dugundji 1966].

If I is a *nilpotent* ideal, i.e., $I^k = (0)$ for some $k > 0$, then the I-adic topology on R is discrete, so R is automatically complete with respect to the I-adic topology.

Even if R is Noetherian, it is quite possible that $\cap I^n \neq (0)$. However, if R is a Noetherian domain and I is any (proper) ideal, or if R is semi-local and I is contained in the Jacobson radical of R, or if R is local and I is any (proper) ideal, then $\cap I^n = (0)$.

Suppose R is a ring, I is an ideal of R, and T is the completion of R in the I-adic topology. Then R is dense in T, topologically. If J is an ideal of R, let $\overline{J}$ be the closure of J. Then $\overline{J}$ is an ideal of T and $\overline{J} \cap R = J$. In particular, $\overline{I}$ is an ideal of T, and T is complete in the $\overline{I}$-adic topology.

If u is a unit in R and $a \in I$, then $u + a$ is invertible in T. $\overline{I}$ is contained in the Jacobson radical of T; thus the only maximal ideals of R which survive in T are those which contain I.

If R is Noetherian, so is T. Assume R is Noetherian. Then T is Cohen–Macaulay or regular if and only if R is Cohen–Macaulay or regular, respectively. If T is a domain, so is R, but the converse of this is false.

Suppose furthermore that R is Noetherian and $I = (a_1, a_2, \ldots, a_m)$. Then the structure of $T = \overline{R}$ is easy to describe: T is isomorphic to $R[[X_1, \ldots, X_m]] / (X_1 - a_1, \ldots, X_m - a_m)$, and $\dim(T)$ = the supremum of $\mathrm{ht}(P)$ for all prime ideals P containing I. Even though R is Noetherian, we may have $\dim(T) < \dim(R)$, and LO may fail for the pair R, T. However, GD always holds.

Suppose R is a semi-local ring and I is the Jacobson radical of R. Then T is isomorphic to $R_1 \oplus R_2 \oplus \ldots \oplus R_m$, where R has m maximal ideals $M_1, \ldots, M_m$, and R_i is the completion of R in the M_i-adic topology. Other information about T can be gathered by studying the rings R_i, each of which is a complete local ring.

This brings us to the case in which R is a local ring with maximal ideal M and T is the completion of R in the M-adic topology. T is a local ring, with maximal ideal $\overline{M}$, and $\overline{M} \cap R = M$. Since GD holds, we find that $\dim(T) = \dim(R)$, and in fact LO also holds for R, T in this case.

If T is an integrally closed domain, or a Krull domain, or a factorial domain, then R is also, respectively, integrally closed, Krull, or factorial. On the other hand, it is possible for R to be an integrally closed domain while it completion T may have nilpotent elements, or T may have more than one prime ideal of height 0. It is also possible for R to be factorial while T is not factorial.

If R is any ring, $R[[X]]$ is the completion of $R[X]$ in the (X)-adic topology. This illustrates the fact that a non-Noetherian ring might be finite-dimensional while its completion (in some I-adic topology) might be infinite-dimensional.

Also, in some cases, $R[X]$ might be integrally closed while $R[[X]]$ might fail to be integrally closed.

Ⓖ. Ⓜiscellaneous Ⓒonstructions

In addition to the types of ring extensions described in the preceding sections, there are various other frequently encountered constructions. Suppose S_1 and S_2 are two rings, each containing R as a subring. Then S_1 and S_2 are R-modules, and we can form their tensor product $T = S_1 \otimes_R S_2$. T is an R-module, and it can also be considered as a ring, with the multiplication defined on basis elements by $(s \otimes t)(s' \otimes t') = ss' \otimes tt'$.

This definition obviously encompasses many possibilities. T can be considered as the "smallest" ring containing both S_1 and S_2 as subrings. (Or, rather, T contains homomorphic images of S_1 and S_2, by the mappings $s_1 \rightarrow s_1 \otimes 1$ and $s_2 \rightarrow 1 \otimes s_2$.) It is not essential that R actually be a subring of S_1 and S_2; it is sufficient that S_1 and S_2 be R-algebras. (In the language of category theory, T is the coproduct of S_1 and S_2 in the category of R-algebras.)

We are concerned here with the observation that even if R, S_1, and S_2 are very "well-behaved" rings, T may have a more complicated structure. Even if R, S_1, and S_2 are fields, T may have zero-divisors, or even nilpotent elements. In particular, the tensor product of two fields is not generally a field. If R is a field and S_1 and S_2 are affine rings over R, then T is certainly an affine ring.

In algebraic geometry one frequently deals with rings which are localizations of affine rings over a given field K. If R is such a ring, it contains K as a subfield; and it is often necessary to enlarge the subfield K to some larger field L. The ring $T = R \otimes_K L$ is then of interest. An integrally closed domain R is said to be *geometrically normal* ("normal" is, in one of its too many usages, a synonym for "integrally closed") if T is an integrally closed domain whenever L is a finite algebraic extension of K. (It is actually necessary to consider only finite inseparable algebraic extensions.) Likewise, a regular domain R is said to be *geometrically regular* if T is a regular domain whenever L is a finite algebraic extension of K. (Again, only inseparable extensions need be considered.)[18]

As the existence of the terms geometric normality and geometric regularity implies, there are integrally closed domains (in characteristics other than 0)

---

[18] One account of geometric normality and geometric regularity is in [Murthy 1976].

which are not geometrically normal, and there are regular domains (not of characteristic 0) which are not geometrically regular.

It is worth noticing that many rings can be constructed in a variety of ways. For instance, if K is a field, then $K[X, Y]$ can be considered as $K[X][Y]$, or as $K[X] \otimes K[Y]$, or as a semigroup algebra over K, or as $K[X, Y, Z]/(Z)$.

Sometimes it is useful to study the union T of an ascending chain $R_1 \subseteq R_2 \subseteq \ldots$ of rings.[19] Almost any ring T which is sufficiently "large" can be represented (nontrivially) as such a union, so there is little one can say about ascent and descent questions that holds in every case. If each R_i is a domain or a field, then T is clearly a domain or a field, respectively.

Suppose each ring R_i is a domain. If also each R_i is integrally closed, Prüfer, Bézout, or a valuation domain, then T has the same property, respectively. On the other hand, it is possible for all of the rings R_i to be Noetherian while T is not Noetherian, or for T to be Noetherian while none of the rings R_i are. The union of an ascending chain of factorial domains is not necessarily factorial, or even completely integrally closed.

If one assumes a bit more about the relationships between the rings, more results are available. The union of a chain of integral extensions is still an integral extension. If, for $i > 0$, R_i is a factorial domain and the irreducible elements of R_i are also irreducible in R_{i+1}, then the union T is also a factorial domain.

Suppose $\dim(R_i) \leq n$ for all $i > 0$; then $\dim(T) \leq n$. If the set $\{\dim(R_i)\}$ is unbounded, then T may be of any dimension, finite or infinite.

Instead of a union of rings, let us consider an intersection. Suppose we have a descending chain of rings, $R_1 \supseteq R_2 \supseteq R_3 \supseteq \ldots$. Let T be the intersection of the rings R_i. Any ring T can be represented as such an intersection (by using suitable polynomial rings in infinitely many variables).

Clearly the intersection of any family (not necessarily a chain) of domains (or fields) is a domain (or a field, respectively). If each of the domains R_i is integrally closed, completely integrally closed, or a valuation domain, then T also has the same property, respectively. On the other hand, it is possible for each ring R_i to be Noetherian while the intersection T is not Noetherian.

If each ring R_i is n-dimensional, then $\dim(T)$ may be n or less than n. If $n > 0$, then $\dim(T) > n$ is possible. Even if R_i is integral over R_{i+1}, for each $i > 0$,

---

[19] Some of the material on unions and intersections of chains of rings can be found in [Bourbaki 1972], and some of it is folklore.

dim(T) may still be less than n. (That is, R_i might not be integral over T, for any i.)

Without further information on the rings R_i and their relationships (perhaps involving LO or INC), there is not a great deal more that can be said about arbitrary intersections and unions.

Intersections of rings (not necessarily in a chain) occur in several contexts. Any Krull domain is the intersection of all of its localizations at prime ideals of height 1. Another instance is the following situation.

Let R be a ring and let G be a group of automorphisms of R. If $g \in G$, let R_g be the subring of R consisting of all the elements of R left fixed by g. Let R_G be the subring consisting of all elements left fixed by every element of G. Clearly then R_G is just the intersection of all the rings R_g.

Suppose $g \in G$ is of finite order, i.e., g^m = the identity automorphism for some positive integer m. Then R is integral over R_g. Why? Let $t \in R$, and let $a_i = g^i(t)$, for $1 \leqslant i \leqslant m$. Then t is a root of the monic polynomial $(X - a_1) \cdot (X - a_2) \dots (X - a_m)$ over R. In fact, all of the coefficients of this polynomial are in R_g, since they are symmetric functions of $a_1, \dots, a_m$. Therefore, t is integral over R_g, so R is integral over R_g.

It follows that R is integral over R_G if G is finite. Under appropriate conditions, R_G may inherit various properties from R. For instance, if R is integrally closed and G is finite, then R_G is also integrally closed. It is possible for R to be Noetherian while R_G is not. This whole situation has been studied in detail.[20]

[20] See [Bourbaki 1972, Chapter V], or [Gilmer 1972a, p. 114 ff.].

4

Homological Dimension

A. Various Types of Modules

The subject of homological algebra is best approached by dealing with modules over not necessarily commutative rings. The structure of the rings is of great importance, since homological algebra of vector spaces over a field collapses into triviality. The following discussion is quite limited.[1]

Let R be a ring and let M be an R-module. M is said to be a *free* R-module if it is the direct sum of copies of R. That is, a finitely generated free R-module M has a basis $a_1, a_2, \ldots, a_m$ such that $M = a_1 R + a_2 R + \ldots + a_m R$, and $a_i R \cap a_j R = (0)$ for $i \neq j$. A map f of M into another R-module N is thus determined completely by the values $f(a_1), f(a_2), \ldots, f(a_m)$, which may be chosed arbitrarily. This sort of mapping property is characteristic of free algebraic objects in general.

It is well known that if R is a field, then any R-module whatsoever is free, i.e., any vector space has a basis. If R is any other ring, not a field, then there are R-modules which are not free.

---

[1] General references for this chapter are [Kaplansky 1974a], [Kaplansky 1974b], and [Rotman 1970]. See [Nichols 1972].

Consider the following property of a module F:

(*) Whenever M is an R-module with a submodule N, and g is a mapping from F into M/N, there is a mapping h from F to M such that the composite $F \to M \to M/N$ is precisely g.

A module that satisfies (*) is called *projective*. Free R-modules are projective, as can be seen by considering the fate of each basis element, but not all projective modules are free. An equivalent definition is that M is a projective R-module if and only if M is a direct summand of a free R-module.

It is of considerable interest to ask for which rings all projective modules are free, or all finitely generated projective modules are free, or all ideals are projective, or all finitely generated ideals are projective. An imposing quantity of algebra, commutative and otherwise, has resulted from studying these questions.

If R is a principal ideal domain or a local ring, then every projective module is free.

If R is a domain, then an ideal I is projective if and only if I is invertible (and therefore finitely generated). A ring R is called *hereditary* if every ideal is projective, and is called *semi-hereditary* if every finitely generated ideal is projective. It follows that a hereditary domain is precisely a Dedekind domain, and a semi-hereditary domain is precisely a Prüfer domain. (These concepts are of more interest in the noncommutative case, referring to, say, left ideals; or in the presence of zero-divisors.)

Suppose R is a ring such that every R-module is projective. Then R is precisely the direct sum of a finite family of fields; if R is a domain, it must be a field.

In category theory, one often studies the dual of a statement or concept. We therefore have the following property of a module E: (dual to (*))

(**) Whenever M is an R-module with a submodule N, and g is a mapping of N into E, there is a mapping h from M into E such that the composite $N \to M \to E$ is precisely g.

A module that satisfies property (**) is called *injective*. Explicit examples of injective modules are few and far between, save in more or less trivial cases. In particular, R itself is not usually injective as an R-module. (If R is injective over itself and Noetherian, it is called *quasi-Frobenius*.)

If E is an R-module and for any non-zero-divisor $r \in R$ and any $x \in E$ there is an element $y \in E$ with $ry = x$, then E is said to be a *divisible R*-module. Notice that r may be a zero-divisor on E; we may have $ry = x$ and $rx = 0$.

Injective R-modules are divisible, but the converse is quite false.

If R is a domain, an R-module M is said to be *torsion-free* if for $r \in R$ and $x \in M$, $rx = 0$ implies $r = 0$ or $x = 0$. (The concept clearly makes sense only if R is a domain.) If R is a domain, then a torsion-free R-module M is injective if and only if it is divisible.

If R is a domain, then R is Dedekind if and only if every divisible R-module is injective.

It is somewhat interesting to notice that while finitely generated projective modules are commonplace (R itself, for one), one does not often encounter finitely generated injective modules. Of course, if R is a field, then every R-module is injective.

Let F be an R-module and suppose that we have an exact sequence $0 \to L \to M \to N \to 0$ of R-modules. F is said to be a *flat* R-module if for every such sequence of modules, the induced sequence $0 \to F \otimes_R L \to F \otimes_R M \to F \otimes_R N \to 0$ is exact. F is said to be *faithfully flat* if it is flat and $F \otimes_R P = 0$ implies $P = 0$.

R itself is clearly faithfully flat. Any projective module is flat, any any free module is faithfully flat, but flat modules are not, in general, projective or faithfully flat.

Suppose R is a domain. Then a flat module M must be torsion-free. R is Prüfer if and only if every torsion-free module is flat. For any ring R and any flat R-module M, an element a of R is a zero-divisor on R if and only if it is a zero-divisor on M.

Suppose R is a ring and T is an extension of R. If T is flat as an R-module, then GD holds for the pair R, T.[2] It is therefore useful to know when T is flat, for different types of extensions.

If R is a ring, and S is a multiplicatively closed subset of R, then R_S is a flat R-module, but may not be faithfully flat. If X is an indeterminate over R, then $R[X]$ is a free R-module, and therefore faithfully flat. If R is a Noetherian ring, I is an ideal of R, and T is the completion of R in the I-adic topology, then T is a flat R-module. It follows that if R is Noetherian, $R[[X]]$ is flat as an $R[X]$-module, and for that matter as an R-module.

A considerable amount of work has gone into studying when various types of extensions, such as $R[X]/I$, yield flat R-modules.[3]

[2] See [Matsumura 1970, p. 33] for a proof.
[3] See [Vasconcelos 1970] for material on this.

$\mathbb{B}_\square$ $\mathbb{P}$rojective $\mathbb{D}$imension

Let R be any ring. If M is any R-module, M can be represented as a homomorphic image of a free R-module. This is very easy: Let F be the free R-module having as its basis elements f_m , where m ranges over all elements of M. There is an obvious mapping of F onto M.

This gives us an exact sequence $0 \to K \to F \to M \to 0$, where K is the kernel of the mapping $F \to M$. However, K can also be represented in the same way, giving us another such sequence. These sequences can be pieced together to produce an exact sequence of the form $\ldots \to F_3 \to F_2 \to F_1 \to M \to 0$, where each of the modules F_i is free. This is called a *free resolution* of M. It is one way to obtain a *projective resolution* of M; a projective resolution is an exact sequence of the form $\ldots \to P_3 \to P_2 \to P_1 \to M \to 0$, where each of the modules P_i is projective.

Since infinite sequences are awkward to handle, it is convenient if the projective resolution of M terminates after some finite number of steps, i.e., if there is such a resolution having $P_i = 0$ for $i > n$, for some positive integer n. If this is the case, and M admits a resolution of length n, but no shorter resolution, then n is the *projective dimension* of M, denoted pd(M), or $\text{pd}_R(M)$ if we want to specify the ring R. If M is not of any finite projective dimension, we have pd(M) = ∞.

Notice that pd(M) = 0 if and only if M is projective, and pd(M) = 1 if and only if M is of the form P/Q, where both P and Q are projective and M itself is not projective.

Suppose we have a short exact sequence $0 \to A \to B \to C \to 0$ of R-modules, each of finite projective dimension. Then we have the following relationships:

(i)   If pd(B) > pd(A), then pd(C) = pd(B).
(ii)   If pd(B) < pd(A), then pd(C) = pd(A) + 1.
(iii) If pd(B) = pd(A), then pd(C) $\leqslant$ pd(B) + 1.

All of these possibilities can occur.

It is quite possible for a single ring to have modules of various projective dimensions. We define the *global dimension* of R to be GD(R) = the supremum of pd(M) for all R-modules M. We have Gd(R) = 0 if and only if R is the direct sum of finitely many fields. If R is a domain, Gd(R) $\leqslant$ 1 if and only if R is Dedekind.

The most important question concerning Gd(R) seems to be not its actual value, but whether it is finite or infinite. If R is a local ring, Gd(R) is finite if and only if R is regular. (This is one way to show that any localization of a regular ring is regular, since $Gd(R) \geqslant Gd(R_P)$ for any prime ideal P.)[4] Specifically, if R is local and $Gd(R)$ is finite, then $Gd(R) = \dim(R)$.

In defining $Gd(R)$, it is sufficient to consider only finitely generated modules, or even only modules of the form R/I for all ideals I.

$\mathbb{C}_{\square}$ $\lvert$njective $\lvert\mathbb{D}\rangle$imension

Let R be any ring and let M be an R-module. Just as M can be expressed as a homomorphic image of a free module, it can also be embedded as a submodule of an injective module. In fact, there is a module $E(M)$, which is injective, which contains M as a submodule, and which is the smallest such module in the sense that any injective module containing M must contain an isomorphic copy of $E(M)$. $E(M)$ is called the *injective envelope* of M; proving its existence is not trivial.[5]

This gives us an exact sequence $0 \to M \to E(M) \to E(M)/M \to 0$. However, $E(M)/M$ can be embedded in $E(E(M)/M)$, giving us another such sequence. These sequences can be pieced together to obtain an exact sequence $0 \to E_1 \to E_2 \to E_3 \to \dots$, where each module E_i is injective. This is called an injective resolution of M.

Just as in the case of projective dimension, we define the *injective dimension* of M, $id(M)$, to be n if $E_i = 0$ for all $i > n$, for some positive integer n, and at least one such injective resolution of M exists (and no shorter resolution exists); and $id(M) = \infty$ if no such n exists.

Global dimension was defined in terms of $pd(M)$, and it is clear that instead a concept of global dimension could be defined in terms of $id(M)$. However, it is a beautiful and interesting fact that whether $Gd(R)$ is defined in terms of projective dimension or injective dimension, the same number is obtained.

Suppose R is a local ring. R is said to be *Gorenstein* if $id(R)$ is finite, where R is considered as a module over itself. Since $id(R) \leqslant Gd(R)$, it is clear that a regular local ring is Gorenstein. It is also true that Gorenstein local rings are Cohen–Macaulay, but there are local rings which are Cohen–Macaulay but

---

[4] For a proof, see [Kaplansky 1974a] and [Kaplansky 1974b]. This is an instance of a recurrent phenomenon: there are various ring-theoretic facts concerning regular local rings which appear to require the use of at least some homological machinery to prove.

[5] See [Rotman 1970] for an existence proof.

not Gorenstein, and there are local rings which are Gorenstein but not regular (and not even domains). If R is Gorenstein, then $id(R) = \dim(R)$.

Suppose R is a Noetherian ring. Then R is said to be *Gorenstein* if R_M is Gorenstein for every maximal ideal M of R. Again, regular rings are Gorenstein and Gorenstein rings are Cohen–Macaulay. If R is a Gorenstein ring, then so is any localization, so is $R[X]$, and so is $R[[X]]$. If R is Gorenstein and I is an ideal of R generated by an R-sequence $a_1, a_2, \ldots, a_m$, then R/I is also Gorenstein. This last fact is useful in generating examples of Gorenstein and non-Gorenstein rings.

If R is an arbitrary ring and M is an R-module, $pd(M)$ and $id(M)$ may not have much in common. For instance, if R is considered as a module over itself, we have $pd(R) = 0$, but $id(R)$ might be anything. If R is a local ring and we require $id(R) = 0$, then in fact R must be Gorenstein (by definition, since 0 is finite), and 0-dimensional. Thus, the maximal ideal of R must be nilpotent. Quasi-Frobenius rings must be Gorenstein and 0-dimensional.

There are more theories of homological dimension available than we have outlined in this and the previous sections. For instance, "weak dimension" is defined in terms of flat modules.

5

Rings without 1

A. What is a Ring without 1?

In all of the preceding chapters, it has been assumed routinely that all of the rings had multiplicative identity elements, i.e., it has been assumed that if R is a ring, R must contain an element 1 such that for any $x \in R$, $1 \cdot x = x$.

It is by no means essential to make this assumption. Rings which arise in the study of algebraic geometry are invariably rings with 1; however, there are plenty of rings without 1, such as infinite direct sums of rings, or some rings of functions. There are various weaker assumptions that can be made. Consider the following properties:

(i) R has 1.
(ii)   For each $x \in R$, there is a $y \in R$ with $xy = x$.
(iii) $R^2 = R$.

Property (i) implies (ii), and (ii) implies (iii), and neither converse is true. There are other properties also of interest.[1]

---

[1] See [Gilmer 1972a, pp. 31–32] for a longer list of not all equivalent statements relating to whether a ring has 1.

If R is a ring with 1, there is a large body of theory available discussing the structure of R. For rings without 1, much less can be said. One way to study a ring R without 1 is to embed it in a larger ring T which has 1. There are several ways available to do this.

Suppose R is any ring. We define a new ring T_1 to be $R \oplus \mathbf{Z}$, with operations defined by $(x, m) + (y, n) = (x + y, m + n)$ and $(x, m)(y, n) = (xy + nx + my, mn)$. Then T_1 is a ring with identity $(0, 1)$ and T_1 is integral over R. (If R is an algebra over K for some ring K with 1, a similar construction yields a ring $T_1{}'$ which contains R and K.)[2]

If R is a ring in which not every element is a zero-divisor, then R has a total quotient ring R_S, where S is the set of all non-zero-divisors of R. Let T_2 be the subring of R_S generated by R and 1.

If R is a domain,[3] let T_3 be the intersection of all nontrivial localizations of R (considered as subrings of the quotient field of R).

These are three of the different ways to adjoin 1 to a ring R, and the three rings T_1, T_2, T_3 may be all distinct. For instance, T_1 is always of characteristic 0, and T_3 may be considerably larger than T_2.

B. Disadvantages of a Ring without 1

There are many theorems of commutative ring theory which fail to hold true if the rings involved are not assumed to have 1. Here are a few samples:

If R is a ring with 1 in which every prime ideal is principal, then every ideal is principal. If R is a ring without 1, it is possible that every prime ideal may be principal, but not every ideal.

Suppose that R is a subring of T, and T is integral over R. If R has 1, then INC holds for the pair R, T. If R does not have 1, this may fail, and $\dim(R)$ may not be $\dim(T)$.

If R is a principal ideal domain with 1, then it is necessarily a Dedekind domain, so that any nonzero ideal is the product of (finitely many) prime ideals. However, if R is a domain without 1, it is possible to have every ideal principal without every ideal being the product of prime ideals.

---

[2] See [Arnold and Gilmer 1974b] for a discussion of $\dim(R)$ and $\dim(T_1)$.

[3] In this chapter we will use the word "domain" to mean a commutative ring without zero-divisors, i.e., a domain might not have an identity.

If R is a Noetherian ring with 1, then $R[X]$ is also Noetherian. If R does not have 1, then $R[X]$ is never Noetherian. (It is a rather curious fact that if $R[X]$ is Noetherian, then R is Noetherian and has 1.)[4]

Since any finite domain is a field, any ring without 1 must either have zero-divisors or be infinite in cardinality. It is possible to have a ring R without 1 in which there are no prime ideals other than R itself. In such a ring, there may be maximal ideals which are not prime.

[4] This theorem is the title and purpose of [Gilmer 1967a].

Miscellaneous Topics

A. G-domains

Let R be an integral domain with quotient field K. R is said to be a *G-domain* if it satisfies any of the (equivalent) conditions.[1]

(i) K is finitely generated as a ring over R.

(ii) K is generated by one element over R, i.e., $K = R[u^{-1}]$ for some non-zero $u \in R$.

(iii) There is a nonzero element $u \in R$ such that any nonzero prime ideal of R contains u.

(iv) $R[X]$ has a maximal ideal M satisfying $M \cap R = (0)$. (In this case, $K \cong R[X]/M$.)

It follows from the definition above, especially (i), that any overring (in particular, any localization) of a G-domain is a G-domain. On the other hand, $R[X]$ is never a G-domain for any ring R.

---

[1] See [Kaplansky 1974b, pp. 12–19] and [Gilmer 1972a, p. 374 ff.] for material on G-domains and Hilbert rings.

Any finite-dimensional valuation domain is a G-domain, and so is any 1-dimensional semi-quasi-local domain. It follows that G-domains may fail to be integrally closed. A Noetherian domain R is a G-domain if and only if it is semi-local and 1-dimensional.

A prime ideal P of a ring R is called a *G-ideal* if it satisfies any of the (equivalent) conditions:

(i) R/P is a G-domain.
(ii) $P = M \cap R$ for some maximal ideal M of $R[X]$.
(iii) For some nonzero element $u \in R$, P is maximal among all ideals with respect to excluding the powers of u.

Clearly, any maximal ideal is a G-ideal, but it is definitely not true that every G-ideal is maximal. (This fact adds much to the complexity of commutative ring theory.) In fact, a G-ideal may be of any height; in a finite-dimensional valuation domain every prime ideal is a G-ideal.

Any radical ideal (hence, any prime ideal) is the intersection of all G-ideals containing it.

B. Hilbert Rings

A ring R is said to be a *Hilbert ring* if it satisfies any of the (equivalent) conditions:

(i)   Every G-ideal is maximal.
(ii) Every prime ideal is an intersection of maximal ideals.
(iii) Every radical ideal is an intersection of maximal ideals.
(iv) Every G-ideal is an intersection of maximal ideals.
(v)   If M is a maximal ideal in $R[X]$, then $M \cap R$ is a maximal ideal in R.
(vi) The Jacobson radical of R/P is (0) for every prime ideal P of R.

A Noetherian ring R is a Hilbert ring if and only if for each prime ideal P $\dim(R/P) = 1$, P is contained in infinitely many maximal ideals.

If R is a Hilbert ring having only finitely many maximal ideals, then these are all of the nonzero prime ideals of R. (Or, more ponderously, any semi-quasi-local Hilbert ring must be 0-dimensional.) If R is Hilbert and $\dim(R) > 0$, then R must have infinitely many maximal ideals.

In general, localizations of a Hilbert ring are usually not Hilbert, but any homomorphic image of a Hilbert ring is Hilbert. A ring R is Hilbert if and only if $R[X]$ is Hilbert. Thus, any affine ring is Hilbert.

Only fields are both Hilbert rings and G-domains.

The concept of a Hilbert ring is also of interest for noncommutative rings. (For instance, conditions (ii) and (vi) are of interest.)

C. von Neumann Regular Rings

Let R be a ring. An element $a \in R$ is said to be *von Neumann regular* (which we shall abbreviate to VNR) if there is an element $x \in R$ with $axa = a$. A ring is said to be *VNR* if each of its elements is VNR.[2]

The concept of von Neumann regularity is of great interest, but more for noncommutative than for commutative rings. The term honors J. von Neumann, who used this concept for studying rings of operators.

For commutative rings, von Neumann regularity is a very strong property. The commutative ring R is VNR if and only if

(i) R is 0-dimensional and $\text{rad}(R) = (0)$, or
(ii) R_M is a field for every maximal ideal M of R, or
(iii) R is Bézout and each principal ideal is generated by an idempotent element, or
(iv)   every R-module is flat.

If R is a VNR ring, then R is its own total quotient ring, i.e., each element $a \in R$ is either a unit ($ax = 1$) or a zero-divisor ($a(xa - 1) = 0$). A VNR domain is a field, and any Noetherian VNR ring is the direct sum of a finite number of fields.

If R is VNR, the Jacobson radical of R is (0). This is a partial converse to (i).

If R is VNR, then any localization of R is also VNR. The direct sum of a finite number of VNR rings is VNR.

If R is VNR, then every finitely generated R-module is projective.

---

[2] See [Gilmer 1972a, p. 111 ff.] or [Kaplansky 1974a, p. 110 ff.] for information about von Neumann regularity.

Part II
The Examples

Example 1

Let $R = K[X_1, X_2, \ldots]$, where K is a field. R is called the free algebra *on countably many generators over K.*

(a) R is not Noetherian. Specifically, the ideal M generated by $X_1, X_2, \ldots$ is not generated by any finite set of generators.

(b) R is of infinite Krull dimension. The ideal M above is maximal, and we have $(0) \subset (X_1) \subset (X_1, X_2) \subset \ldots \subset M$, so $\mathrm{ht}(M) = \infty$.

(c) R is a factorial domain, so it satisfies the ascending chain condition on principal ideals.

(d) If Q is a finitely generated prime ideal of R, the generators of Q are polynomials in finitely many of the X_i, say $X_1, \ldots, X_m$. Then R_Q is a localization of $K(X_{m+1}, X_{m+2}, \ldots)[X_1, \ldots, X_m]$, so R_Q is a regular local ring. Specifically, $\mathrm{ht}(Q)$ is finite.

(e) R is not catenary, since there are noncatenary rings which are countably generated over a field, and any such ring is a homomorphic image of R, for some field K. See Example 28.

(f) R is not Hilbert for reasons exactly like those given in (e).

(g) For variants of this example, instead of the field K we could use **Z**. Instead of countably many generators, we might have uncountably many. Any ring is a homomorphic image of an appropriate free algebra over **Z**.

Example 2

Let $R = \mathbf{Z}[2X, 2X^2, 2X^3, \ldots]$, a subring of $\mathbf{Z}[X]$.

(a) R is not Noetherian. Specifically, the ideal P generated by $2X, 2X^2, \ldots$ is not finitely generated.

(b) Consider the ideals $A = (2X)$ and $B = (2X^2)$. The ideal $A \cap B$ is generated by $4X^2, 4X^3, \ldots$, and is not finitely generated. Thus, even though A and B are principal ideals, $A \cap B$ is not even finitely generated. (It is, of course, divisorial.) In a factorial domain, the intersection of two principal ideals must be principal; therefore R is not factorial.

(c) The maximal ideal $M = (2, 2X, 2X^2, \ldots)$ is minimal over the principal ideal (2). We have $(0) \subset P \subset M$, so $\mathrm{ht}(M) = 2$, and the principal ideal theorem fails for R.

(d) R_M is a 2-dimensional valuation domain, and $R = R_M \cap \mathbf{Z}[X]$, so R is integrally closed. If N is any other maximal ideal of R, then $\frac{1}{2} \in R_N$, so R_N is a localization of $R[\frac{1}{2}] = \mathbf{Z}[\frac{1}{2}, X]$, which is 2-dimensional. Therefore R is 2-dimensional.

(e) R is not completely integrally closed since $2X^n \in R$, and X is in the quotient field of R, but $X \notin R$.

(f) This example is from [Gilmer 1972a, Exercise 21].

Example 3

Let $R = K[Y, XY, X^2 Y, X^3 Y, \ldots]$, a subring of $K[X, Y]$, where K is a field.

(a) This ring has almost precisely the same properties as the ring of Example 2. Thus, the maximal ideal $M = (Y, XY, X^2 Y, \ldots)$ is minimal over the principal ideal (Y), i.e., minimal among all prime ideals containing (Y), so the principal ideal theorem fails for R, and R is neither Noetherian nor factorial. R is integrally closed and 2-dimensional, and is not completely integrally closed. R is a nonaffine subring of $K[X, Y]$.

(b) One difference between this example and the ring in Example 2 is that here char(K) may be nonzero. Both rings are special cases of the following construction:

Let T_0 be a principal ideal domain. Let t be a principal prime element of T_0 and let $T = T_0[tX, t^2 X, t^3 X, \ldots]$. (It is interesting to notice that with all of the other properties, T satisfies the ascending chain condition on principal ideals.)

(c) This example is due to G. Evans, and is from [Kaplansky 1974b, p. 114, Exercise 8].

Example 4

Let K be a field. Let $R = K[X_{11}, X_{21}, X_{22}, \ldots, X_{n1}, X_{n2}, \ldots, X_{nn}, \ldots]$. For each $n > 0$, let $P_n = (X_{n1}, \ldots, X_{nn})$. Let S be the complement of $P_1 \cup P_2 \cup$

$P_3 \cup \ldots$. *Let $T = R_S$.*

(a) T is actually a Noetherian ring. (Notice that each of its prime ideals is finitely generated.)

(b) Nonetheless, $\dim(T) = \infty$. The ideals P_n are each maximal, with $\mathrm{ht}(P_n) = n$, and these are all of the maximal ideals of T.

(c) T is a regular domain, since each T_{P_n} is a regular local domain.

(d) T is factorial since R is.

(e) T is not Hilbert. Consider the prime ideal $J = (X_{22})$. We have $\dim(T/J) = 1$ and T/J is local; thus J is contained in only one maximal ideal, and is not an intersection of maximal ideals.

(f) This example is from [Nagata 1962, p. 203].

□ □ □ □ □

Example 5

Let $R = K[X, Y]$, where K is a field

(a) The ring R is a 2-dimensional, regular, factorial Hilbert domain.

(b) Let $M = (X, Y)$. M is a maximal ideal, so $M^2 = (X^2, XY, Y^2)$ is an M-primary ideal. However, M^2 is reducible: $M^2 = (X, Y^2) \cap (X^2, Y)$.

(c) The ideal $I = (X^2, Y)$ is M-primary, since $M^2 \subseteq I \subseteq M$; but I is not a power of a prime ideal.

(d) The ideal $J = (X^2, XY)$ is not primary, but $\mathrm{rad}(J) = (X)$ is a prime ideal.

(e) One primary decomposition of J is $J = (X) \cap M^2$. Here (X) is prime and M^2 is M-primary. Thus the associated primes of J are (X) and M: M is an embedded prime of J and (X) is an isolated prime.

(f) J has many other decompositions. For instance, $J = I \cap (X)$. If $c \in K$, $J = (Y + cX, X^2) \cap (X)$, where $(Y + cX, X^2)$ is M-primary. Thus the component of J belonging to the embedded prime M is not unique.

(g) This example is from [Northcott 1953, pp. 29–30], [van der Waerden 1970, vol. 2, pp. 130–131], and [Zariski and Samuel 1958, pp. 154, 208, and 210].

□ □ □ □ □

Example 6

Let $R = K[X, Y, Z]$, where K is a field.

(a) The ring R is a 3-dimensional, regular, factorial Hilbert domain.

(b) Let $M = (X, Y, Z)$. M is a maximal ideal. The ideal $I = (X, Y^2, Z^3)$ is M-primary, but it is not a power of a prime ideal.

(c) Let $P = (f_1, f_2, f_3)$, where $f_1 = Y^2 - XZ$, $f_2 = YZ - X^3$, and $f_3 = Z^2 - X^2 Y$. P is a prime ideal, and not maximal, since $P \subset M$. For an element g of R, $g \in P$ if and only if $g(t^3, t^4, t^5) = 0$.

(d) P^2 is not P-primary. To see this, notice that $X(X^5 - 3X^2 YZ + XY^3) = f_2^2 - f_1 f_3 \in P^2$. For P^2 to be P-primary, we would have to have $X^n \in P^2$ for some n, or $X^5 - 3X^2 YZ + XY^3 \in P^2$, both of which fail. Of course, P^2 does have a P-primary component.

(e) This example is from [Northcott 1953, pp. 29–30].

Example 7

Let $R = K[X, Y, Z]/(XY - Z^2)$, where K is a field. Let x, y, z denote the images of X, Y, Z, respectively.

(a) R is Cohen–Macaulay, 2-dimensional, integrally closed, and a domain.

(b) R is not regular. To see this, let $M = (x, y, z)$. M is a maximal ideal of height 2. If R were regular, R_M would have to be factorial; but consider the ideal $P = (x, z)$. P is prime, of height 1, contained in M, and P is not principal in R, nor is P_M principal in R_M. Therefore R_M is not factorial, and R is not regular (nor factorial).

(c) P^2 is not P-primary, since $xy = z^2 \in P^2$, but $x \notin P^2$ and $y^n \notin P^2$ for any $n > 0$.

(d) We have $P^2 = (x^2, xz, xy) = (x) \cap (x^2, xy, y)$, where (x) is P-primary and (x^2, xy, y) is M-primary. Here M is an embedded prime of P^2.

(e) A somewhat less arduous way to show that R is not regular is to observe that $XY - Z^2$ is in the square of a maximal ideal, namely (X, Y, Z). If T is a regular ring and $t \in T$ is in the square of a maximal ideal of T, then $T/(t)$ will not be regular.

(f) This example is from [Zariski and Samuel 1958, p. 154].

Example 8

*Let R = **Z**[X].*

(a) R is a 2-dimensional, regular, factorial Hilbert domain.

(b) Let $I = (X^2, 2X)$. I is not primary, since $2X \in I$, $X \notin I$, and $2^n \notin I$ for any $n > 0$. However, $(X)^2 \subseteq I \subseteq (X)$, and (X) is a prime ideal; thus rad(I) is a prime ideal, even though I is not primary.

(c) Let $J = (4, 2X, X^2)$. J is M-primary for the maximal ideal $M = (2, X)$, but J is reducible: $J = (4, X) \cap (2, X^2)$.

(d) Let $L = (9, 3X + 3)$. One primary decomposition for L is $L = (3) \cap (9, X + 1)$, where (3) is a prime ideal and $(9, X + 1)$ is N-primary for the maximal ideal $N = (3, X + 1)$. Thus the associated prime ideals of L are (3) and $(3, X + 1)$, respectively an isolated and an embedded prime ideal of L.

(e) This example is from [van der Waerden 1970, vol. 2, pp. 122, 126, and 130].

□ □ □ □ □

Example 9

*Let R = **Z**[3X, X^2, X^3].*

(a) R is Noetherian and 2-dimensional.

(b) R is not integrally closed; its integral closure is $\mathbf{Z}[X]$.

(c) Let $P = (3X, X^2, X^3)$. P is a prime ideal, but not maximal. P^2 is not primary, since $9X^2 \in P^2$, but $X^2 \notin P^2$ and $9^n \notin P^2$ for any $n > 0$.

(d) R is not Cohen–Macaulay. If $M = (3, 3X, X^2, X^3)$, then the single element 3 by itself is a maximal R-sequence in R_M. (For any $b \in M$, $3Xb = 3d$ for some $d \in M$.)

(e) This example is from [van der Waerden 1970, vol. 2, p. 123].

□ □ □ □ □

Example 10

Let R be any integral domain, not a field. Let P be a nonzero prime ideal of R.

(a) The sequence $0 \to P \to R \to R/P \to 0$ is exact. We have $\mathrm{Ass}(R) = \{(0)\}$, and $\mathrm{Ass}(R/P) = \{P\}$. Thus the inclusion $\mathrm{Ass}(R) \subseteq \mathrm{Ass}(P) \cup \mathrm{Ass}(R/P)$ is proper.

(b) This example is from [Matsumura 1970, p. 51].

□ □ □ □ □

Example 11

Let $R = K[[X^2, X^3]]$, where K is a field.

(a) R is a 1-dimensional local domain.

(b) R is not integrally closed. Its integral closure is $K[[X]]$, a Noetherian valuation domain.

(c) R is a G-domain. Specifically, $R[X^{-1}]$ is the quotient field of R, namely $K((X))$.

(d) This example is from [Kaplansky 1974b, p. 41, Exercise 5].

□ □ □ □ □

Example 12

Let $R = K[X, Y]/(Y^2 - X^3 - X^2)$, where K is a field.

(a) R is a 1-dimensional, Noetherian, Hilbert domain.

(b) R is not integrally closed. Its integral closure is $T = K[y/x]$, where x, y are the images of X, Y, respectively. Notice that $(y/x)^2 = x + 1$. T is a principal ideal domain.

(c) This example is from [Dieudonne 1967, p. 12].

□ □ □ □ □

Example 13

Let $R = \mathbf{Z} + 2i\mathbf{Z}$ ($i = \sqrt{-1}$).

(a) R is a 1-dimensional, Noetherian, Hilbert domain.

(b) R is not integrally closed. Its integral closure is $T = \mathbf{Z} + i\mathbf{Z}$. T is a principal ideal domain.

(c) This example is from [Kaplansky 1974b, p. 41, Exercise 4].

□ □ □ □ □

Example 14

Let $R = K[X, Y]/(X^2 - Y^3)$, where K is a field. Let x, y be the images of X, Y, respectively.

(a) R is a 1-dimensional, Noetherian, Hilbert domain.

(b) R is not integrally closed. Its integral closure is $T = K[x/y]$. Notice that $(x/y)^2 = y$. T is a principal ideal domain.

(c) This example is from [Zariski and Samuel 1958, p. 262].

□ □ □ □ □

Example 15

Let $R = \mathbf{Z}[\sqrt{5}]$.

(a) R is a 1-dimensional, Noetherian, Hilbert domain.

(b) R is not integrally closed. Its integral closure is $T = \mathbf{Z}[t]$, where $t = (1 + \sqrt{5})/2$. Notice that $t^2 - t - 1 = 0$. T is a principal ideal domain.

(c) Notice that T is the integral closure of $\mathbf{Z}$ in $\mathbf{Q}(\sqrt{5})$ and is not just $\mathbf{Z}(\sqrt{5})$. However, $\mathbf{Q}(\sqrt{5}) = \mathbf{Q}(t)$.

□ □ □ □ □

Example 16

For each $n > 0$, let $R_n = K[X, Y/X^n]$, where K is a field. We have $R_1 \subseteq R_2 \subseteq R_3 \subseteq \ldots$. Let $R = $ the union of the R_n.

(a) For each $n > 0$, R_n is isomorphic to $K[X, Y]$, and is therefore a 2-dimensional, regular, factorial, Hilbert domain. In particular, R_n is completely integrally closed.

(b) $R = K[X, Y/X, Y/X^2, Y/X^3, \ldots]$.

(c) R is not completely integrally closed. Its quotient field is $K(X, Y)$. We have $Y \in R$ and $X^{-1} \notin R$, but $Y(X^{-1})^n \in R$ for every $n > 0$.

(d) R is 2-dimensional and integrally closed, since R_n is 2-dimensional and integrally closed for each $n > 0$.

(e) R is therefore not factorial nor Noetherian.

(f) Note the similarities between this example and Example 3.

(g) This example is from [Gilmer 1972a, p. 145, Exercise 12].

Example 17

Let $R = K[X, Y, Z, U, V]/(XY + Z^2 + U^2 + V^2)$, where K is a field such that char$(K) \neq 2$. Let x, y, z, u, v be the images of X, Y, Z, U, V, respectively. Let $M = (x, y, z, u, v)$. M is a maximal ideal. Let $T = R_M$.

(a) T is a 4-dimensional local domain.

(b) T is not regular since $XY + Z^2 + U^2 + V^2 \in (X, Y, Z, U, V)^2$, but T is Cohen–Macaulay.

(c) T is factorial. To see this, notice that $T/(x)$ is a domain, so x is a principal prime. Now $T[x^{-1}]$ is isomorphic to $K[X, Z, U, V]_N[X^{-1}]$, where $N = (X, Z, U, V)$, which is factorial. Since T is Noetherian, x is a principal prime, and $T[x^{-1}]$ is factorial, T is also factorial.

(d) Notice that R and T would not even be domains if char$(K) = 2$.

Example 18

For each $n > 0$, let $R_n = \mathbf{R}[X_i X_j \text{ for } 1 \leqslant i, j \leqslant n]$. Let $s = X_1^2 + X_2^2 + \ldots + X_n^2$. Let $D_n = R_n[s^{-1}]$.

(a) R_n is an n-dimensional, integrally closed, Cohen–Macaulay domain.

(b) D_n is also an n-dimensional, integrally closed, Cohen–Macaulay domain.

(c) Let A_n = the ideal of D_n generated by the monomials $X_1 X_j$ for $1 \leqslant j \leqslant n$. Now A_n is invertible, since $A_n^2 = (X_1^2)$. (A_n is also prime.)

(d) A_n has no basis of fewer than n elements. The proof is topological: nonzero functions on the n-sphere are induced from nonzero elements of D_n.

(e) Since A_n is a nonprincipal prime ideal of height 1 in D_n, D_n (and therefore also R_n) is not factorial, for $n > 1$.

(f) This example is from [Gilmer 1969a].

Example 19

Let K be the field of two elements. Let $R = K[X, Y]/(X^2, XY, Y^2)$. Let x, y be the images of X, Y, respectively.

(a) R is a 0-dimensional, local, Cohen–Macaulay ring, with maximal ideal $M = (x, y)$. (R is thus trivially a Hilbert ring.)

(b) We have $M \subseteq (x) \cup (x + y) \cup (y)$, but M is not contained in any one of these ideals. This phenomenon (an ideal contained in a finite union of ideals but not contained in any one of them) cannot happen if all of the ideals are prime.

(c) Instead of using the field of two elements, we could have used any finite field and devised a similar example.

(d) This example is from [Gilmer 1972a, p. 43].

Example 20

Let $\mathbf{Z}$ be the ring of integers.

(a) Let P_i range over all (proper) prime ideals of $\mathbf{Z}$. Let $I = (5)$. The ideal $(3, I)$ is all of $\mathbf{Z}$, so $(3, I) \not\subseteq$ the union of the P_i, but for each $i \in I$, $3 + i \in$ the union of the P_i.

(b) This example is from [Kaplansky 1974b, p. 103, Exercise 17].

Example 21

Let $R = \mathbf{Z}_{(p)}$, where p is a fixed prime integer.

(a) R is a Noetherian valuation domain, with maximal ideal (p). R is trivially a G-domain, as $R[p^{-1}] = \mathbf{Q}$.

(b) Let $M = \mathbf{Q}$ = the quotient field of R. M is a flat, injective, faithful R-module.

(c) M is not finitely generated as an R-module.

(d) We have $(p) \cdot M = M$. Thus the Nakayama lemma can fail if the module involved is not finitely generated. Notice that $\cap (p)^n \cdot M = M$.

(e) This example is from [Matsumura 1970, p. 22].

Example 22

Let $R = K[X, Y, Z]$, where K is a field.

(a) R is a 3-dimensional, regular, factorial, Hilbert domain.

(b) The sequence X, $Y(1 - X)$, $Z(1 - X)$ is an R-sequence of (maximal) length 3.

(c) The sequence $Y(1 - X)$, $Z(1 - X)$, X is not an R-sequence. Thus the order of an R-sequence cannot be changed at will.

(d) Let $I = (Y(1 - X), Z(1 - X))$ and let $J = (X)$. We have $Gr(I, R) = 1$, $Gr(J) = 1$, and $Gr(I + J, R) = 3$.

(e) This example is from [Kaplansky 1974b, p. 102, Exercise 7].

Example 23

Let $T = R[X]$, where R is a Noetherian valuation domain with maximal ideal (p) (for example, R might be $\mathbf{Z}_{(3)}$).

(a) T is a 2-dimensional, regular, factorial domain.

(b) The sequence X, p is a maximal R-sequence, of length 2. The sequence $pX - 1$ is also a maximal R-sequence, of length 1. Thus there may be maximal

R-sequences of different lengths.

(c) Notice that the ideals (X, p) and $(pX - 1)$ are both maximal, of heights 2 and 1, respectively.

(d) *R* is a G-domain and not a field, so it is not Hilbert, and therefore *T* is not Hilbert either.

(e) This example is from [Kaplansky 1974b, p. 103, Exercise 16].

Example 24

Let $R = K[X, Y]/(X^2, XY)$, where K is a field. Let x, y be the images of X, Y, respectively.

(a) *R* is a 1-dimensional, Noetherian, Hilbert ring.

(b) Let $M = (x, y)$. *M* is a maximal ideal, of height 1 and $\mathrm{Gr}(M, R) = 0$. Thus *R* is not Cohen–Macaulay.

(c) On the other hand, $R/(x)$ is isomorphic to $K[Y]$, which certainly is Cohen–Macaulay.

(d) This example is from [Kaplansky 1974b, p. 102, Exercise 6].

Example 25

Let $R = K[X, Y, Z]/(X^n + Y^n + Z^n)$, where K is a field of characteristic p and n is a positive integer not divisible by p. Let x, y, z be the images of X, Y, Z, respectively. Let $S = R[U, V]$. Let $T = K[xU, yU, zU, xV, yV, zV]$. Thus T is a subring of S.

(a) *R* is a 2-dimensional, integrally closed, Cohen–Macaulay domain.

(b) *S* is a 4-dimensional, integrally closed, Cohen–Macaulay domain.

(c) *T* is a Noetherian integrally closed domain.

(d) *T* is not Cohen–Macaulay. Specifically, if *M* is the ideal (xU, yU, zU, xV, yV, zV), then *M* is maximal and T_M is not Cohen–Macaulay. The sequence xU, yV is a maximal *R*-sequence in *M*, and in fact $\mathrm{Gr}(M, T) = 2$.

(e) T_M is 3-dimensional. To see this, notice that T is an affine ring, and so catenary. Therefore T_M is catenary. Notice that $(0) \subset (xU, yU, zU) \subset (xU, yU, zU, xV, yV - zV) \subset M$ is a saturated chain of prime ideals contained in M. Therefore, $ht(M) = 3$ and $\dim(T_M) = 3$.

(f) T_M is obviously local.

(g) In fact, T is also 3-dimensional, and T_Q is regular for any maximal ideal $Q \neq M$.

(h) Thus T_M is a 3-dimensional, local, integrally closed domain which is not Cohen–Macaulay.

(i) This example is from [Murthy 1976, vol. 2, p. 37].

Example 26

For each $n > 0$, let $R_n = K[X, Y_1, Y_2, \ldots, Y_n]/(XY_1, XY_2, \ldots, XY_n)$, where K is a field.

(a) For each $n > 0$, R_n is n-dimensional and Noetherian.

(b) For $n \geqslant 2$, inside the ring R_n, let $M = (X, Y_1, Y_2, \ldots, Y_n)$. Then $(Y_1, \ldots, Y_n) \subset M$ and $(X) \subset (X, Y_1) \subset (X, Y_1, Y_2) \subset \ldots \subset (X, Y_1, Y_2, \ldots, Y_{n-1}) \subset M$ are both saturated chains of prime ideals contained in M, of lengths 1 and n respectively. Notice that the prime ideals (X) and $(Y_1, Y_2, \ldots, Y_n)$ are minimal.

(c) In spite of this, R_n is catenary, for all $n > 0$, since R_n is an affine ring.

Example 27

Let $R = K(Y)[[X]]$, where K is a field. Let T be the subring consisting of all elements of R whose constant terms are in K: that is, $T = K + X \cdot K(Y)[[X]]$.

(a) T is a 1-dimensional, integrally closed, quasi-local domain.

(b) T is neither Noetherian nor a valuation domain.

(c) $T[Z]$ is 3-dimensional. To see this, consider the following diagram of mappings, each of which corresponds to a prime ideal, its kernel.

$$T[Z] \xrightarrow{\ Z\ =\ Y\ } K[Y] + X \cdot K(Y)[[X]]$$

$$\downarrow Z = 0 \qquad\qquad\qquad \downarrow X = 0$$

$$T \xrightarrow{\ X\ =\ 0\ } K \xleftarrow{\ Y\ =\ 0\ } K[Y]$$

One side of this diagram is of length 3, corresponding to a chain of prime ideals of length 3, so $\dim(T[Z]) = 3$. (Of course, since T is 1-dimensional, $T[Z]$ can be of dimension at most 3.)

(d) Also, the left side of the diagram is of length 2, corresponding to a chain of prime ideals of length 2. The diagram commutes, so both chains of prime ideals are contained in the same maximal ideal. Both chains are saturated; therefore, $T[Z]$ is not catenary.

(e) $T[Z]$ is integrally closed. If we want, we can make it quasi-local as well, retaining the interesting features listed above.

(f) Neither T nor $T[Z]$ is completely integrally closed, since X and Y are in the quotient field of T and $XY^n \in T$ for all n, yet $Y \notin T$.

(g) Any domain satisfying (a) and (b) satisfies (c).

(h) The statement in (g) is from [Kaplansky 1974b, p. 42, Exercise 18].

□ □ □ □ □

Example 28

Our construction of this example will be in several parts.

Part I. *Let $W_0 = K[X, Y]$, where K is a field, and let $I = (X - 1, Y)$. I is a maximal ideal of W_0. Let $W = (W_0)_I$.*

(a) W is a 2-dimensional, regular, local ring with maximal ideal I_I.

(b) Notice that K is mapped isomorphically onto the residue field W/I_I.

Part II. *Let V be a Noetherian valuation domain, containing $K[X, Y]$ and contained in $K(X, Y)$, such that if J is the maximal ideal of V, then $J \cap K[X, Y] = (X, Y)$, and K is mapped isomorphically onto the residue field V/J.*

(c) One way to obtain such a V is as follows: Let s be an element of $K[[t]]$ which is transcendental over $K(t)$ and which is of the form $s = t + a_2 t^2 + a_3 t^3 + \dots$. Map W_0 isomorphically into $K[[t]]$ by sending X to t and Y to

s (and sending K identically onto itself). This map can be extended to the quotient fields of W_0 and $K[[t]]$. V will be the inverse image of $K[[t]]$.

Part III. *Let* $T = V \cap W$.

(d) T has just two maximal ideals, M and N, such that $T_M = V$ and $T_N = W$.

(e) Since T is semi-quasi-local and each localization of T at a maximal ideal is Noetherian, T is Noetherian.

(f) T is integrally closed.

(g) T is 2-dimensional. Specifically, $\mathrm{ht}(M) = 1$ and $\mathrm{ht}(N) = 2$.

(h) Notice that $K \subseteq T$, and K is mapped isomorphically onto both residue fields T/M and T/N.

Part IV. *Let* $R = K + (M \cap N)$.

(i) R is a quasi-local ring with maximal ideal $M \cap N$.

(j) R is 2-dimensional. To see this, let P be a prime ideal of T of height 1, contained in N. That is, $(0) \subset P \subset N$. In R, we have $(0) \subset M \cap P \subset M \cap N$, so $\dim(R) \geqslant 2$.

(k) T is finitely generated as an R-module. In fact, $T = R[e]$ for any element e of N that is not in M. To see this, observe that $T = K + N = M + N = R + N$. As an R-module, $T/R \cong (R + N)/R \cong N/N \cap R = N/M \cap N \cong (M + N)/M = T/M \cong K$; thus T is generated as an R-module by R and one more element.

(l) In particular, T is integral over R. T is actually the integral closure of R. Since T is Noetherian and a finitely generated R-module, R is also Noetherian. This reaffirms $\dim(R) = \dim(T) = 2$.

(m) $R[Z]$ is not catenary. Why? Let Q be the kernel of the map of $R[Z]$ onto T defined by sending Z to e (for some particular choice of e in (k) above). Q is a principal prime ideal in $R[Z]$, so $\mathrm{ht}(Q) = 1$. Let M_0 be the inverse image, under this mapping, of M. M_0 is a maximal ideal in $R[Z]$.

However, there can be no prime ideals lying between Q and M_0 in $R[Z]$, since the image of M_0 in T is just M, which is of height 1. On the other hand, $R[Z]/M_0$ is precisely T/M, so $M_0 \cap R$ must be precisely $M \cap N$. Therefore, we have $(0) \subset (M \cap P) \cdot R[Z] \subset (M \cap N) \cdot R[Z] \subset M_0$, and $(0) \subset Q \subset M_0$, both saturated chains, of different lengths.

(n) Notice that the choice of e is fairly arbitrary, so there are many ideals Q available. Likewise, there are many choices available for P. Thus, there are many saturated chains of each length.

(o) Notice that if K is countable, $R[Z]$ is also. Then $R[Z]$ is a homomorphic image of the ring of Example 1. Any finite ring is necessarily catenary, as a homomorphic image of $\mathbf{Z}[X_1, X_2, \dots, X_n]$, for a large enough n.

(p) This example is from [Nagata 1962, pp. 203–205]; also see [Matsumura 1970, pp. 87–88] and [Kaplansky 1976, pp. 19–21].

□ □ □ □ □

Example 29

This example is a more elaborate version of the preceding example. Let $W_0 = K[X, Y_1, \ldots, Y_m, Z_1, \ldots, Z_r]$, where K is a field. (We allow $m = 0$ here; but $r \geqslant 1$.) Let $I = (X - 1, Y_1, \ldots, Y_m, Z_1, \ldots, Z_r)$. I is a maximal ideal of W_0. Let $W = (W_0)_I$.

Notice that W is an $(m + r + 1)$-dimensional, regular, local domain, and that K is mapped isomorphically onto the residue field of W/I_I.

Let K_1 be the quotient field of W. Let V be an $(m + 1)$-dimensional, regular, local domain such that $W_0 \subseteq V \subseteq K_1$, and if J is the maximal ideal of V, then $J \cap W_0 = I$, and K is mapped isomorphically onto the residue field V/J.

(One way to obtain V is to pick algebraically independent elements s_1, $s_2, \ldots, s_r$ in $K[[X]]$, such that each s_i is transcendental over $K(X)$ and is of the form $s_i = X + a_2 X^2 + a_3 X^3 + \ldots$. We then map W_0 isomorphically into $K[[X]][Y_1, \ldots, Y_m]$ by sending K identically onto K, X to X, Y_i to Y_i, and Z_i to s_i. Extend the map to the quotient fields and let V be the inverse image of $K[[X]][Y_1, \ldots, Y_m]$.)

Let $T = W \cap V$. T has two maximal ideals, M and N, such that $T_M = V$ and $T_N = W$. Let $R = K + (M \cap N)$.

(a) T is an $(m + r + 1)$-dimensional regular domain. Specifically, $\mathrm{ht}(N) = m + r + 1$ and $\mathrm{ht}(M) = m + 1$.

(b) R is a local domain with maximal ideal $M \cap N$ and $\dim(R) = m + r + 1$.

(c) T is a finitely generated R-module. In fact, $T = R[e]$, for any element e of N that is not in M.

(d) Let U be an indeterminate over K. The domain $R[U]$ is not catenary. Why? Let Q be the kernel of the map of $R[U]$ onto T defined by sending U to e. Let M_0 be the inverse image of M under this mapping. Let $(0) \subset P_1 \subset P_2 \subset \ldots \subset P_m \subset M$ be a saturated chain of prime ideals in T, and for each i, let $P_{i,0}$ be the inverse image of P_i. Then $(0) \subset Q \subset P_{1,0} \subset \ldots \subset P_{m,0} \subset M_0$ is a saturated chain of prime ideals of $R[U]$, of length $m + 2$. On the other hand, M_0 contains $(M \cap N) \cdot R[U]$, which is of height $m + r + 1$. Thus $R[U]$ admits two saturated chains of prime ideals lying between (0) and M_0, of lengths

$m + r + 2$ and $m + 2$.

(e) Notice that in the case $m = 0$ and $r = 1$, i.e., if we omit the Y_i throughout the discussion and have only one Z, then we are left with Example 28.

(f) This example is from [Nagata 1962, pp. 203–205].

□ □ □ □ □

Example 30

Let K be a field and for each $n > 0$, let $K_n = K(X_1, \ldots, X_n)$.

For each n, $m > 0$, let $R_{m,n} = K + Y_1 K_n[[Y_1]] + Y_2 K_{2n}((Y_1))[[Y_2]] + Y_3 K_{3n}((Y_1, Y_2))[[Y_3]] + \ldots + Y_m K_{mn}((Y_1, \ldots, Y_{m-1}))[[Y_m]]$. Let $T_{m,n} = R[Z_1, Z_2, \ldots, Z_n]$.

(a) $R_{m,n}$ is an m-dimensional quasi-local domain.

(b) $R_{m,n}$ is integrally closed but not completely integrally closed, and is therefore not Noetherian; nor is it a valuation domain.

(c) $T_{m,n}$ is of dimension $(m + 1)(n + 1) - 1$ and is not catenary. Let M_0 be the maximal ideal of $R_{m,n}$ (which consists of all elements having zero constant terms). Then $M = (M_0, Z_1, \ldots, Z_n)$ is a maximal ideal of $T_{m,n}$, and is of height $(m + 1)(n + 1) - 1$. T contains saturated chains of prime ideals lying between (0) and M of lengths $(m + 1)(n + 1) - 1$ and $m + n$.

(d) Let us consider the case $m = 1$, $n = 2$, and the following diagram of mappings:

$$
\begin{array}{ccc}
T_{1,2} & \xrightarrow{\;\;Z_2 = X_2\;\;} & K[Z_1, X_2] + Y_1 K_2[[Y_1]][Z_1] \\
\Big\downarrow{\scriptstyle Z_2 = 0} & & \Big\downarrow{\scriptstyle Z_1 = X_1} \\
R_{1,2}[Z_1] & & K[X_1, X_2] + Y_1 K_2[[Y_1]] \\
\Big\downarrow{\scriptstyle Z_1 = 0} & & \Big\downarrow{\scriptstyle Y_1 = 0} \\
R_{1,2} & \xrightarrow{\scriptstyle Y_1 = 0} K \xleftarrow{\scriptstyle X_1 = 0} K[X_1] \xleftarrow{\scriptstyle X_2 = 0} K[X_1, X_2]
\end{array}
$$

These mappings should indicate the kind of arithmetic maneuvering that must occur to construct a long chain of prime ideals.

(e) Lastly, let us consider the case $m = 2$, $n = 1$. (The case $m = n = 1$ is Example 27.)

$$T_{2,1} \xrightarrow{\;Z_1 = X_2\;} K[X_2] + Y_1 K_1[[Y_1]][X_2] + Y_2 K_2((Y_1))[[Y_2]]$$

$$\downarrow {\scriptstyle Z_1 = 0} \qquad\qquad\qquad\qquad\qquad\qquad\qquad\qquad \downarrow {\scriptstyle Y_2 = 0}$$

$$R_{2,1} \qquad\qquad\qquad\qquad\qquad K[X_2] + Y_1 K_1[[Y_1]][X_2]$$

$$\downarrow {\scriptstyle Y_2 = 0} \qquad\qquad\qquad\qquad\qquad\qquad\qquad\qquad \downarrow {\scriptstyle X_2 = X_1}$$

$$K + Y_1 K_1[[Y_1]] \qquad\qquad\qquad K[X_1] + Y_1 K_1[[Y_1]]$$

$$\downarrow {\scriptstyle Y_1 = 0} \qquad\qquad\qquad\qquad\qquad\qquad\qquad\qquad \downarrow {\scriptstyle Y_1 = 0}$$

$$K \xleftarrow{\;\;\;\;\;\;\;\;\;X_1 = 0\;\;\;\;\;\;\;\;\;} K[X_1]$$

Example 31

Let $R = K[X, X^{1/2}, X^{1/3}, X^{1/4}, \ldots]$, *where* K *is a field. Let* Q *be the ideal* $(X, X^{1/2}, X^{1/3}, \ldots)$. Q *is a maximal ideal in* R. *Let* $T = R_Q$.

(a) R is the integral closure of $K[X]$ in the field $K(X, X^{1/2}, X^{1/3}, \ldots)$, which is algebraic over $K(X)$. Therefore, R is a 1-dimensional Prüfer domain, and T is a 1-dimensional valuation domain.

(b) In fact, R is the union of $K[X] \subseteq K[X^{1/2}] \subseteq K[X^{1/6}] \subseteq \ldots \subseteq K[X^{1/n!}] \subseteq \ldots$, an ascending chain of principal ideal domains, so R is actually Bézout.

(c) The value group of T is $G(T) \cong \mathbf{Q}$.

(d) Thus T is not Noetherian, and neither is R.

(e) T is completely integrally closed, but not a Krull domain; and it is not pseudo-principal, since the set $\{x^{p/q} : (p/q)^2 > 2\}$ has no least common multiple.

Example 32

Let $R = \mathbf{Z} + X \cdot \mathbf{Q}[X]$. Let $P = (2, X, X/2, X/3, X/4, \ldots)$. P is a maximal ideal in R. Let $T = R_P$.

(a) R is a 2-dimensional Bézout domain.

(b) T is a 2-dimensional valuation domain.

(c) Neither T nor R is Noetherian, since the ideal $(X, X/2, X/3, \ldots)$ is not finitely generated.

(d) T and R are both integrally closed, but neither is completely integrally closed, since $(\frac{1}{2})^n X \in R$, for all $n > 0$, but $\frac{1}{2} \notin R$ (and likewise for T).

(e) Notice that P is actually a principal ideal, $P = (2)$.

(f) The value group of T is isomorphic to $\mathbf{Z} \oplus \mathbf{Z}$, with the lexicographic ordering $(a, b) > 0$ if and only if $a > 0$ or $a = 0$ and $b > 0$.

Example 33

Let K be a field. Let n be a positive integer. Let G be the group $\mathbf{Z} \oplus \mathbf{Z} \oplus \ldots \oplus \mathbf{Z}$ (n summands). Let G have the lexicographic ordering, i.e., $(a_1, \ldots, a_n) > 0$ if and only if the first of the a_i which is nonzero is > 0. G is thus a linearly ordered group. Let R be a valuation domain having G as its value group and K as its residue field.

(a) R is n-dimensional. To see this, notice that the only convex subgroups of G are of the form $(0) \oplus (0) \oplus \ldots \oplus \mathbf{Z} \oplus \mathbf{Z} \oplus \ldots \oplus \mathbf{Z}$, where m of the n summands are not (0). If H is one of these, H corresponds to a prime ideal P of R, namely, $P =$ the ideal generated by all nonzero elements of R whose images in $G(R)$ do not lie in H. These are all of the prime ideals of R—a chain of length n.

(b) R is Noetherian if and only if $n = 1$.

(c) Since R is a valuation domain, all of its radical ideals are prime. Since $\dim(R)$ is finite, R has the ascending chain condition on prime ideals, hence on radical ideals, but R is not Noetherian (for $n > 1$).

(d) R is completely integrally closed if and only if $n = 1$.

(e) Let $T = R[X]$. T is an $(n + 1)$-dimensional pseudo-Bézout domain.

(f) T is neither Bézout nor, for $n > 1$, Noetherian. To see this, let t be any nonunit in R. Then the ideal (t, X) is not principal. Therefore, T is not Bézout. If T were Noetherian, R would also have to be.

(g) For $n > 1$, T is not pseudo-principal nor factorial, since it is not completely integrally closed.

(h) T contains maximal ideals of various heights. To construct these, let $(0) = P_0, P_1, P_2, \ldots, P_n$ be the prime ideals of R, of heights $0, 1, 2, \ldots, n$, respectively. Let a_i be an element of P_i that is not in P_{i-1}, for $1 \leqslant i \leqslant n$. Then for such i, the ideal $(1 - a_i X, P_{i-1})$ is maximal and of height i. The ideal (P_n, X) is maximal and of height $n + 1$.

(i) In Chapter 2, Section A, the term "discrete valuation domain" was defined. In this example, R is a discrete valuation domain, by that definition. If $G_1, G_2, \ldots, G_n$ are (*any*) subgroups of the additive group of real numbers and $G = G_1 \oplus \ldots \oplus G_n$, with the lexicographic ordering, then we can use the same construction to produce a valuation domain V which will be discrete if and only if G_i is isomorphic to **Z** or is (0) for each i.

□ □ □ □ □

Example 34

Let K be a field and let G be the additive group of real numbers. G is well-known to be a linearly ordered group and complete as a lattice. Let R be a valuation domain having G as its value group and K as its residue field.

(a) R is 1-dimensional but not Noetherian.

(b) Since G is a complete lattice (complete in the sense that bounded sets have glbs and lubs), R is pseudo-principal (and therefore is also completely integrally closed).

(c) R is not factorial (since any 1-dimensional factorial domain is Noetherian).

(d) Let $T = R[X]$. T is pseudo-principal and 2-dimensional , but is not Bézout or factorial.

□ □ □ □ □

Example 35

Let K be a field and let G be a linearly ordered abelian group having no maximal convex subgroups. Let V be a valuation domain having G as its value group *and $L = K(X_1, X_2, \ldots)$ as its residue field.*

(a) One such group G can be obtained as follows:

Let $\mathbf{Z}^-$ be the set of all nonpositive integers. Let G be the set of all functions from $\mathbf{Z}^-$ into $\mathbf{Z}$ having finite support, i.e., if $g \in G$, then $g(m) = 0$ for all but finitely many elements m of $\mathbf{Z}^-$. Define addition in G pointwise; this makes G into an abelian group. Give G the lexicographic ordering: for $g \in G$, $g > 0$ if and only if there is an integer m such that $g(m) > 0$ and $g(n) = 0$ for $n < m$.

G can be pictured as $G = \ldots \oplus \mathbf{Z} \oplus \mathbf{Z} \oplus \mathbf{Z}$. The convex subgroups of G are of the form $H = \ldots \oplus (0) \oplus (0) \oplus \ldots \oplus (0) \oplus \mathbf{Z} \oplus \mathbf{Z} \oplus \ldots \oplus \mathbf{Z}$, so $h \in H$ if there is m_0 such that $h(m) = 0$ for all $m < m_0$. Letting $m_0 = 0, -1, -2, -3, \ldots$, we get an infinitely ascending chain of convex subgroups. Notice that G/H is isomorphic to G, for any of these subgroups H.

(b) V has no prime ideals of height 1, since any such prime ideal would correspond to a maximal proper convex subgroup of G. Vacuously, every prime ideal of height 1 is principal; but V is not factorial.

(c) The complete integral closure of $\overset{.}{V}$ is its quotient field. V is obviously infinite-dimensional.

(d) If P is a nonzero prime ideal of V, then V/P is finite-dimensional. That is, P is of finite *coheight*. The value group of V/P is isomorphic to one of the subgroups H. On the other hand, V_P is infinite-dimensional, with value group isomorphic to G/H for some subgroup H. In fact, V_P is isomorphic to V.

(e) One reason we required L to be the residue field of V was to ensure that V_P would be isomorphic to V.

(f) This example is from [Gilmer 1972a, p. 254, Exercise 20].

□ □ □ □ □

Example 36

Let K be a field and let G be a linearly ordered abelian group having no minimal proper convex subgroups. Let V be a valuation domain having G as its value group and $L = K(X_1, X_2, \ldots)$ as its residue field.

(a) One way to obtain such a group G is as follows:

Let $\mathbf{Z}^+$ be the set of nonnegative integers. Let G be the set of all functions from $\mathbf{Z}^+$ into $\mathbf{Z}$ having finite support, i.e., if $g \in G$, then $g(m) = 0$ for all but finitely many elements m of $\mathbf{Z}^+$. Define addition in G pointwise; this makes

G into an abelian group. Give G the lexicographic ordering: for $g \in G$, $g > 0$ if and only if there is an integer m such that $g(m) > 0$ and $g(n) = 0$ for $n < m$.

G can be pictured as $G = \mathbf{Z} \oplus \mathbf{Z} \oplus \mathbf{Z} \oplus \ldots$. The convex subgroups of G are of the form $H = (0) \oplus (0) \oplus \ldots \oplus (0) \oplus \mathbf{Z} \oplus \mathbf{Z} \oplus \ldots$. Notice that H is isomorphic to G, for any of these subgroups H.

(b) V has no prime ideals lying just below its maximal ideal, i.e., none of co-height 1, since such a prime ideal would correspond to a minimal proper convex subgroup of G. V is obviously infinite-dimensional.

(c) Let P be a nonzero, nonmaximal prime ideal of V. V_P is finite-dimensional; that is, P is of finite height. The value group of V_P is isomorphic to G/H for one of the subgroups H. On the other hand, V/P is infinite-dimensional, with value group isomorphic to H. In fact, V_P is isomorphic to V.

(d) Let P_1 be the unique prime ideal of V of height 1. V_{P_1} is the complete integral closure of V.

(e) Compare this example to Example 35.

Example 37

Let $R = \mathbf{Z}[\sqrt{-5}]$, whose quotient field is $K = \mathbf{Q}(\sqrt{-5})$. Let $T = R + X \cdot K[[X]]$.

(a) R is the integral closure of $\mathbf{Z}$ in K, so R is a Dedekind domain.

(b) T is a Prüfer domain. To see this, suppose P is a prime ideal in T. If $P = (0)$, then T_P is a field. If $P \neq (0)$, we must examine $P^* = P \cap R$. If $P^* = (0)$, then $T_P = K[[X]]$. If $P^* \neq (0)$, then R_{P^*} is a Noetherian valuation domain, and $T_P = R_{P^*} + X \cdot K[[X]]$. It is easy to see that T_P is a 2-dimensional valuation domain. This also shows that $\dim(T) = 2$.

(c) T is not Noetherian, since the ideal $(X, X/5, X/25, \ldots)$ is not finitely generated. (Or since T_P is a valuation domain of dimension > 1.)

(d) T is not Bézout, since the ideal $(2, 3 + \sqrt{-5})$ is finitely generated and not principal.

(e) T is not completely integrally closed, since $X(1/5)^n \in T$ for all $n > 0$, but $1/5 \notin T$. (Or since T_P is a valuation domain of dimension > 1.)

(f) This example is from [Gilmer 1972a, p. 305, Exercise 8].

Example 38

Let $R = K[X, X^{1/2}, X^{1/3}, \ldots, Y, Y^{1/2}, Y^{1/3}, \ldots]$, where K is a field. Let $P_1 = (X, X^{1/2}, X^{1/3}, \ldots)$ and let $P_2 = (Y, Y^{1/2}, Y^{1/3}, \ldots)$. Let $S =$ the complement of $P_1 \cup P_2$. Let $T = R_S$.

(a) T is 1-dimensional and semi-quasi-local, with just two maximal ideals M and N, where $T_M = R_{P_1}$ and $T_N = R_{P_2}$.

(b) T is Prüfer, since T_M and T_N are valuation domains. In fact, T is Bézout, since T is also semi-quasi-local.

(c) T is not Noetherian, since neither M nor N is finitely generated.

(d) T is completely integreally closed, since $T = T_M \cap T_N$, and both T_M and T_N are completely integrally closed.

(e) T is not pseudo-principal, since the set $\{X^{m/n} : (m/n)^2 > 2\}$ has no least common multiple.

(f) T is a G-domain, since $T[(XY)^{-1}]$ is a field.

□ □ □ □ □

Example 39

Let $R = C[X_0, X_1, X_2, \ldots]$. Let I be the ideal generated by all of the elements $X_n^2 - X_{n-1}$, for $n \geqslant 1$. Let $T = R/I$, Let $x_0, x_1, x_2, \ldots$ be the images of $X_0, X_1, X_2, \ldots$, respectively.

(a) T is 1-dimensional, since it is the union of 1-dimensional domains $T_m = C[x_0, x_1, \ldots, x_m]$, where $T_0 \subseteq T_1 \subseteq T_2 \subseteq \ldots$. (Or since T is integral over $C[x_0]$.)

(b) T is Bézout, since each of the domains T_m is a principal ideal domain, and therefore Bézout. (Notice that $T_m = C[x_m]$.)

(c) T is not Noetherian, since the ideal $(x_0, x_1, x_2, \ldots)$ is not finitely generated.

(d) In fact, T does not satisfy the ascending chain condition on radical ideals. Specifically, $(x_0 - 1) \subset (x_1 - 1) \subset (x_2 - 1) \subset \ldots$ is a properly ascending chain of radical ideals. Why are these ideals radical? If M is a maximal ideal of T containing $(x_m - 1)$, M is of the form $(x_0 - e_0, x_1 - e_1, \ldots)$, where $e_0 = e_1 = \ldots = e_m = 1$, and for $n > m$, e_n is a 2^{n-m}th root of 1. There are uncountably many such maximal ideals, and $(x_m - 1)$ is the intersection of them. Thus $(x_m - 1)$ is a radical ideal.

(e) T is not almost Dedekind, since if N is the maximal ideal $(x_0, x_1, \ldots)$, T_N is not Noetherian.

(f) T is completely integrally closed, since it is 1-dimensional and Bézout.

(g) Since T is 1-dimensional, Bézout, and not Noetherian, it is not factorial, and therefore does not satisfy the ascending chain condition on principal ideals.

(h) T is not pseudo-principal, since the set $\{x_0^{m/n} : n$ is a power of 2 and $(m/n)^2 > \frac{1}{2} \}$ does not have a least common multiple.

(i) This example was suggested by C. Weibel.

Example 40

Let K be the field of all algebraic numbers, i.e., the algebraic closure of the field $\mathbf{Q}$. Let R be the integral closure of $\mathbf{Z}$ in K.

(a) Since R is integral over $\mathbf{Z}$, it is clearly 1-dimensional.

(b) R is not Noetherian, since, for instance, the ideal generated by all elements $7^{m/n}$, where $(m/n)^2 > 5$, is not finitely generated.

(c) R is Bézout. (It is easy to prove that R is Prüfer; that R is Bézout is somewhat more difficult.) Since R is 1-dimensional, it is therefore also completely integrally closed.

(d) R is not pseudo-principal, since, for instance, the elements mentioned in (b) have no greatest common divisor.

Example 41

Let L be the field of all meromorphic functions from $\mathbf{C}$ into itself. For simplicity's sake, let us assume that two such functions are equal if they differ only by removable discontinuities. Let R be the ring of all entire functions from $\mathbf{C}$ into itself, i.e., $f \in R$ iff $f \in L$ and f has no poles.

(a) If g is an entire function having no zeroes, the $1/g$ is also entire, so g is a unit in R. Thus, the zeroes of an entire function are of interest. For each $g \in R$, let $Z(g)$ be the set of zeroes of g.

(b) For each $a \in \mathbf{C}$, let P_a = the set of all $g \in R$ such that $g(a) = 0$. P_a is a maximal ideal of R, of height 1. These are not all the maximal ideals of R.

(c) If $g \in R$ and $g \neq 0$, $Z(g)$ is a countable, closed, discrete subset of $\mathbf{C}$. For each $a \in Z(g)$, the order of a as a zero of g is $n_g(a)$ = the largest integer n for which $g \in P_a^n$. Conversely, given a countable, closed, discrete subset W of $\mathbf{C}$, and given a positive integer $m(a)$ for each $a \in W$, then from the Weierstrass theorem of complex analysis, there is an entire function g such that $Z(g) = W$ and $n_g(a) = m(a)$ for each $a \in W$.

One consequence of this is that the quotient field of R is L, which is not an obvious fact (except to analysts).

(d) Suppose $g, h \in R$, both nonzero. We can find $u \in R$ such that $Z(u) = Z(g) \cap Z(h)$ and $n_u(a)$ = the lesser of $n_g(a)$ and $n_h(a)$, for each $a \in Z(u)$. If g and h have no zeroes in common, u has no zeroes and is a unit. It can be shown, using the Mittag-Leffler theorem of complex analysis, that $u \in (g, h)$, so $R = (g/u, h/u)$, and $(u) = (g, h)$.

Thus R is Bézout and u = g.c.d.(g, h).

(e) R is also pseudo-principal. If we have a set S of nonzero elements g of R, we can let W = the intersection of all the sets $Z(g)$, and let $m(a)$ = the infimum of $n_g(a)$ for all g in S, for each $a \in W$. Then the Weierstrass theorem assures us of a greatest common divisor of S.

(f) R is actually infinite-dimensional. This is not at all a trivial fact. What is easy to see is that R must have prime ideals other than the maximal ideals P_a. For each $n > 0$, let f_n be an entire function such that $Z(f_n) = \{n, n + 1, n + 2, \ldots\}$, and $n + k$ is a zero of f_n of order n for each $k \geqslant 0$. The ideal $I = (f_1, f_2, f_3, \ldots)$ is not contained in any of the ideals P_a. I can be enlarged to a maximal ideal N which is not P_a for any $a \in \mathbf{C}$.

(g) Let S be the multiplicatively closed subset of R consisting of all $f \in R$ for which $Z(f)$ is finite. Then R_S is not completely integrally closed. To see this, let $g \in R$ have zeroes at $2, 3, 4, \ldots$, of orders $1, 2, 3, \ldots$, respectively. Let $h \in L$ have poles at $1, 2, 3, \ldots$, each of order 1. Then gh^n has poles at $1, 2, \ldots, n$, of orders $n, \ldots, 2, 1$; so $gh^n \in R_S$, but $h \notin R_S$.

Since there are entire functions g for which $Z(g)$ is infinite, R_S is not all of L. However, L is the complete integral closure of R_S.

(h) Thus we have a domain which is pseudo-principal, but for which some localization is not pseudo-principal, nor even completely integrally closed.

(i) This example is from [Gilmer 1972a, pp. 146–148, Exercises 16–21] and from [Bourbaki 1972, pp. 356–357]. See [Mott 1974] for an account of the infinite Krull dimension of R.

Example 42

Let D be a principal ideal domain and let K be its quotient field. (For instance, we might have $D = \mathbf{Z}$ and $K = \mathbf{Q}$.) Let $R = D + Y \cdot K(X)[[Y]]$.

(a) R is 2-dimensional and integrally closed.

(b) R is not pseudo-Bézout. Specifically, the elements XY and $X^2 Y$ do not have a least common multiple in R.

(c) The irreducible elements of R are precisely the principal primes from D (up to multiplication by units in R). The units of R are of the form $u + Yg$, where u is a unit in D and $g \in K(X)[[Y]]$.

(d) Notice that the principal primes in D remain principal primes in R.

(e) R is not completely integrally closed, since $YX^n \in R$ for all $n > 0$, but $X \notin R$; and thus R is not Noetherian either.

(f) This example is from [Gilmer 1972a, p. 432, Exercise 18].

Example 43

Let S be a nonempty set of prime integers. Let K be a field. Let A = the set of all nonnegative elements of $\mathbf{Z}_S$, Let $R = K[A]$, the semigroup ring of A over K.

(a) A typical element of R is of the form $y = d_0 X^{a_0} + \ldots + d_n X^{a_n}$, where $d_0, \ldots, d_n \in K$ and $a_0, \ldots, a_n \in A$.

(b) R is Bézout and 1-dimensional. We can write A = the union of an ascending chain of cyclic subsemigroups $A_1 \subseteq A_2 \subseteq \ldots$; and likewise R is the union of an ascending chain of rings $R_1 \subseteq R_2 \subseteq \ldots$, where $R_i = K[A_i]$. Since A_i is cyclic, R_i is a principal ideal domain, for each i, and is therefore 1-dimensional and Bézout. R inherits these properties.

(c) The element X is not a finite product of irreducible elements of R. Specifically, X itself is not irreducible, and neither is X^a for any $a \in A$. If R were Noetherian, it would be a principal ideal domain, and every nonunit element would be a product of irreducible elements. Therefore, R is not Noetherian. (If A contained a least positive element a, R would be Noetherian, $R = K[X^a]$.)

(d) R does not satisfy the ascending chain condition on principal ideals. If $p \in S$, then $(X) \subset (X^{1/p}) \subset (X^{1/p^2}) \subset (X^{1/p^3}) \subset \ldots$ is an infinite properly ascending chain of principal ideals.

(e) This example is from [Gilmer 1972a, p. 80, Exercise 15].

□ □ □ □ □

Example 44

Let $R = \mathbf{Z}[X^2, X^3]$.

(a) R is 2-dimensional, Noetherian, and Hilbert.

(b) R is not integrally closed and is therefore not factorial; its integral closure is $\mathbf{Z}[X]$.

(c) The greatest common divisor of the elements X^2 and X^3 is 1, but their least common multiple does not exist in R (l.c.m.$(X^2, X^3) \neq X^5$ since each of them divides X^6, and X^5 does not divide X^6 in R). The greatest common divisor of X^5 and X^6 does not exist in R, and neither does their least common multiple. All of these l.c.m.s and g.c.d.s do exist, of course, in $\mathbf{Z}[X]$.

(d) This example is from [Gilmer 1972a, p. 76, Exercise 1].

□ □ □ □ □

Example 45

Let $R = \mathbf{Z}[\sqrt{-5}]$. Let $T = R[X]$.

(a) R is the integral closure of $\mathbf{Z}$ in $\mathbf{Q}(\sqrt{-5})$, so R is a Dedekind domain.

(b) The greatest common divisor of 3 and $k = 2 + \sqrt{-5}$ is 1, but the ideal $(3, k)$ is not principal; i.e., 1 cannot be written as $3a + kb$ for any $a, b \in R$. The elements 9 and $3k$ do not have a greatest common divisor in R.

(c) Thus R is not factorial, so it is not a principal ideal domain. R is a 1-dimensional π-domain.

(d) T is a 2-dimensional regular domain. Since R is not factorial, neither is T. Thus T is an example of a 2-dimensional π-domain which is not factorial.

(e) This example is from [Gilmer 1972a, p. 77, Exercise 4].

□ □ □ □ □

Example 46

Let $G_0 = \mathbf{Z} \oplus \mathbf{Z}$. Let G be the subgroup consisting of all elements (a, b) with $a + b$ even. Give G the partial order defined by $(a, b) \geqslant 0$ if and only if $a \geqslant 0$ and $b \geqslant 0$.

(a) G is a partially ordered abelian group.

(b) G is torsion-free.

(c) G is directed; given two elements (a, b) and (c, d), the element $(2 \cdot \max(a, c), 2 \cdot \max(b, d))$ is clearly an upper bound for both.

(d) G is not the group of divisibility of any domain R. In order to examine the reason for this, we need some terminology.

Suppose K is a field and H is a partially ordered abelian group. A mapping $w: K \rightarrow H \cup \{\infty\}$, where ∞ is defined to be greater than any element of H, is called a *semivaluation* if it satisfies the following:

(i) $w(0) = \infty$.

(ii) $w(ab) = w(a) + w(b)$, for all $a, b \in K^*$.

(iii) $w(-1) = 0$.

(iv)   For all $a, b \in K^*$ with $a \neq -b$, and all $g \in H$, if $w(a), w(b) \geqslant g$, then $w(a + b) \geqslant g$.

(If H is lattice-ordered, condition (iv) simply says that $w(a + b) \geqslant$ the lesser of $w(a)$ and $w(b)$.)

If R is a domain with quotient field K and divisibility group $G(R) = K^*/R^*$, then the mapping $w : K \rightarrow G(R) \cup \{\infty\}$, where $w(a) = a^*$, is a semivaluation, in which $w(t) \geqslant 0$ if and only if $t \in R$. Conversely, if $w : K \rightarrow H \cup \{\infty\}$ is a semi-valuation, then K contains a domain R for which $H = G(R)$.

Now let us consider our group G. Suppose $G \cong G(R)$ for some domain R; let R have quotient field K and let $w : K \rightarrow G \cup \{\infty\}$ be a semivaluation.

G contains the elements $(2, 2)$ and $(3, 1)$. For some $a, b \in R$, $w(a) = (2, 2)$ and $w(b) = (3, 1)$. Let $w(a + b) = (m, n)$. Since $(1, 1)$ and $(2, 0)$ are both smaller than $(2, 2)$ and $(3, 1)$, we must have (m, n) larger than both $(1, 1)$ and $(2, 0)$. Thus $c \geqslant 2$ and $d \geqslant 1$.

If $c = 2$, then $d \geqslant 2$ since $c + d$ must be even and $d \geqslant 1$. If $d = 1$, then $c \geqslant 3$, likewise. In either case, $w(a + b) \geqslant w(a)$ or $w(a + b) \geqslant w(b)$. Hence, either $(a + b) \subseteq (a)$ or $(a + b) \subseteq (b)$, so either $(a) \subseteq (b)$ or $(b) \subseteq (a)$. That is, either a divides b or b divides a, so either $w(a)$ or $w(b)$ is larger. However, $(2, 2)$ and $(3, 1)$ are not comparable. This is a contradiction. Thus G is not $G(R)$ for any domain R.

(e) Of course, any partially ordered group G_0 which is not torsion-free cannot be the group of divisibility of any integrally closed (or even root-closed) domain.

(f) This example is due to Jaffard ([Jaffard 1956]). Also see [Mott 1973], from which this account is borrowed. The proof is due to Paul Hill.

Example 47

Let $R = \mathbf{Z}[X, \sqrt{2X}]$.

(a) R is a 2-dimensional, Noetherian, integrally closed, Hilbert domain. (So it is Cohen–Macaulay.)

(b) Let $P = (X, \sqrt{2X})$. P is a prime ideal of height 1. Since P is not principal, R is not factorial. If $M = (2, X, \sqrt{2X})$, M is a maximal ideal of R, and P_M is not principal in R_M. Therefore, R_M is not factorial, and R is not regular (or a π-domain).

(c) P^2 is not divisorial. Specifically, $P^2 = (2X, X\sqrt{2X}, X^2)$, so $I = P^{-1} = R[\sqrt{2/X}]$, and $I^{-1} = P$. Thus $((P^2)^{-1})^{-1} \neq P^2$, so P^2 is not divisorial. This is in contrast to the 1-dimensional case; in a Dedekind domain, every nonzero ideal is divisorial.

(d) This example is from [Gilmer 1972a, p. 554, Exercise 2].

Example 48

Let $T = K[X_1, X_2, \ldots]$; where K is a field and let $R = K[X_iX_j$ for all i and j ranging over $1, 2, 3, \ldots]$; so R is a subring of T.

(a) T is the ring of Example 1, so it is infinite-dimensional and factorial. In particular, each prime ideal of T which is of height 1 is principal. The ideal $P = (X_1)$ is a prime ideal of T of height 1.

(b) $R = T \cap L$, where $L = K(X_iX_j$ for all i and $j)$. L is a subfield of the quotient field of T, so R is a Krull domain. R is not the integral closure of any Noetherian domain, since R has maximal ideals of infinite height, such as, e.g., $M = (X_iX_j$ for all i and $j)$. Notice that $(P \cap R)^2 = (X_1^2)M$.

(c) T is integral over R, so R is also infinite-dimensional. The prime ideal $P \cap R$ is a prime ideal of R of height 1. However, $P \cap R = (X_1 X_j$ for $j = 1, 2, \ldots)$.

(d) Thus we have a Krull domain R containing a prime ideal P which is of height 1, but which is not finitely generated, much less invertible or principal.

(e) Also see Example 18.

(f) I do not know whether R has any finitely generated nonzero prime ideals. (It seems unlikely.)

(g) This example is from [Eakin and Heinzer 1968].

Example 49

Let $R = K[X, Y, U, V]/(XU - YV)$, where K is a field. Let x, y, u, v be the images of X, Y, U, V, respectively.

(a) R is a 3-dimensional, Noetherian, Hilbert domain.

(b) R is a Krull domain. To see this, let $P = (x, y)$. P is a prime ideal of R of height 1. $R = R_P \cap K[x, y, v/x]$, where R_P is a Noetherian valuation domain and $K[x, y, v/x]$ is a regular domain. Since R is an intersection of Krull domains, it is Krull.

(c) Since P is not principal, R is not factorial. The divisor class group of R is $\mathrm{Cl}(R) \cong \mathbf{Z}$. $\mathrm{Cl}(R)$ is generated by the divisor class of P.

(d) It is worthwhile to notice that if $\mathrm{char}(K) \neq 2$ and $\sqrt{-1} \in K$, R is isomorphic to $K[X, Y, U, V]/(X^2 + Y^2 + U^2 + V^2)$. R is one of the simplest non-factorial domains.

(e) This example is from [Fossum 1973, pp. 65–66].

Example 50

Let R be an arbitrary Krull domain. Let I be a nonempty set. Let $T = R[X_i, Y_i, U_i, V_i$ for all $i \in I]/(X_i U_i - Y_i V_i$ for all $i \in I)$.

(a) T is a Krull domain.

(b) T is not factorial. Its divisor class group is $\text{Cl}(T) \cong \text{Cl}(R) \oplus \mathbf{Z}^I$, where $\mathbf{Z}^I$ is the free abelian group on the set I.

(c) If R is factorial, then $\text{Cl}(T) \cong \mathbf{Z}^I$. If R is a field and I is a singleton set, then this is precisely Example 49.

(d) T is not Noetherian unless R is Noetherian and the set I is finite. T is not finite-dimensional unless $\dim(R)$ is finite and the set I is finite.

(e) This example, like Example 49, is from [Fossum 1973, pp. 65–66]. See also [Claborn 1966]. Using this example, in the case in which R is a field, we can realize any free abelian group as $\text{Cl}(T)$. Of course, T is not a Dedekind domain; to prove that there is a Dedekind domain with the same divisor class group is not easy.

Example 51

Let K be a field and let m be an integer $\geqslant 2$. Let $R = K[X_1, \ldots, X_m, Y_1, \ldots, Y_m]/(X_iY_j - X_jY_i$ for $1 \leqslant i < j \leqslant m)$. Let x_i, y_j be the images of X_i, Y_j, respectively, for all i and j.

(a) R is a Noetherian domain of dimension $m + 1$.

(b) R is a Krull domain.

(c) If $P = (x_1, \ldots, x_m)$, P is a prime ideal of R of height 1. Since P is not principal, R is not factorial. In fact, $\text{Cl}(R) \cong \mathbf{Z}$. $\text{Cl}(R)$ is generated by the divisor class of P.

(d) If $m = 2$, this is precisely Example 49.

(e) This example can be expanded, in the same fashion as Example 49 was expanded to obtain Example 50.

(f) This example is from [Fossum 1973, p. 67].

Example 52

Let K be an algebraically closed field of characteristic $\neq 2$. Let $R = K[X, Y, Z]$. Let $T = R/(X^2 + Y^2 + Z^2)$. Let x, y, z be the images of X, Y, Z, respectively.

(a) T is a 2-dimensional, integrally closed, Cohen–Macaulay, Hilbert domain.

(b) Let $i \in K$, $i^2 = -1$. Let $u = y + iz$, $v = y - iz$. Then $uv = y^2 + z^2$, and $uv + x^2 = 0$. $T = K[u, v, x]$.

(c) Let $P = (u, x)$. P is a prime ideal of T of height 1. Thus T is not factorial. We have $P \subseteq M = (x, u, v)$, and P_M is not principal in T_M, so T_M is not factorial, and T is not regular or a π-domain.

(d) $P^2 = u(x, u, v) = uM$, so $\mathrm{div}(P^2) = ((P^2)^{-1})^{-1} = (u)$, which is principal. Thus $\mathrm{div}(P)$ is of order 2 in $\mathrm{Cl}(T)$. In fact, $\mathrm{Cl}(T) \cong \mathbf{Z}/2\mathbf{Z}$. $\mathrm{Cl}(T)$ is generated by $\mathrm{div}(P)$.

(e) If K is not algebraically closed, T may be factorial. Instead of the polynomial $X^2 + Y^2 + Z^2$, we could use any nondegenerate quadratic form F in R. The precise condition is that $\mathrm{Cl}(T) \cong \mathbf{Z}/2\mathbf{Z}$ if and only if there is a nontrivial solution to $F(X, Y, Z) = 0$ in K. Otherwise, T is factorial.

(f) This example is from [Fossum 1973, p. 51].

Example 53

Let K be a field of characteristic $\neq 2$. Let F be a nondegenerate quadratic form in $K[X, Y, Z, W]$. Let $T = K[X, Y, Z, W]/(F)$.

(a) T is a 3-dimensional, integrally closed, Cohen–Macaulay, Hilbert domain.

(b) The question of the factoriality of T depends on the precise choice of K and F. In general, $\mathrm{Cl}(T)$ is either (0) or isomorphic to $\mathbf{Z}$.

(c) If K is algebraically closed, $T = K[s, t, u, v]$, where $st = uv$, and s, t, u, v are appropriately chosen linear combinations of the images of X, Y, Z, W. This is just Example 49, so $\mathrm{Cl}(T) \cong \mathbf{Z}$.

(d) Let K be the field of real numbers. If $F(X, Y, Z, W) = X^2 + Y^2 + Z^2 + W^2$, then $\mathrm{Cl}(T) \cong \mathbf{Z}$. If $F(X, Y, Z, W) = X^2 + Y^2 + Z^2 - W^2$, then $\mathrm{Cl}(T) = (0)$. If $F(X, Y, Z, W) = X^2 + Y^2 - Z^2 - W^2$, then $\mathrm{Cl}(T) \cong \mathbf{Z}$.

(e) Let K be the field of rational numbers. If $F(X, Y, Z, W) = X^2 + Y^2 + Z^2 + 2W^2$, then $\mathrm{Cl}(T) = (0)$. If $F(X, Y, Z, W) = X^2 + Y^2 + 2Z^2 + 2W^2$, then $\mathrm{Cl}(T) \cong \mathbf{Z}$.

(f) This example is from [Fossum 1973, pp. 52–53].

Example 54

Let K be a field of characteristic $\neq 2$. Let F be a nondegenerate quadratic form in $K[X_1, \ldots, X_n]$. Let $R = K[X_1, \ldots, X_n]/(F)$.

(a) For $n \geqslant 3$, R is an integrally closed, Cohen–Macaulay, Hilbert domain of dimension $n - 1$.

(b) If $n = 2$, R is 1-dimensional, Cohen–Macaulay, and Hilbert, but may not be a domain, unless F is irreducible. If R is a domain, it is not integrally closed.

(c) If $n \geqslant 5$, R is factorial.

(d) The case $n = 3$ is treated in Example 52. The case $n = 4$ is treated in Example 53. If $n = 2$ and R is a domain, its integral closure is a principal ideal domain.

(e) This example is from [Fossum 1973, p. 50].

□ □ □ □ □

Example 55

Let K be either the field of real numbers or the field of complex numbers. Let n be a positive integer. Let $R = K[X_0, \ldots, X_n]/(X_0^2 + \ldots + X_n^2 - 1)$.

(a) R is an n-dimensional, integrally closed, Cohen–Macaulay, Hilbert domain.

(b) Let K be the field of complex numbers. If $n = 1$ or $n \geqslant 3$, R is factorial. If $n = 2$, $Cl(R) \cong \mathbf{Z}$.

(c) Let K be the field of real numbers. If $n \geqslant 2$, R is factorial. If $n = 1$, $Cl(R) \cong \mathbf{Z}/2\mathbf{Z}$.

(d) In either case, if $n = 0$, R is not a domain.

(e) This example is from [Fossum 1973, p. 53].

□ □ □ □ □

Example 56

Let K be a field of characteristic $p > 0$. Let q be a power of p. Let $R = K[X^q, XY, Y^q]$.

(a) R is a 2-dimensional, integrally closed, Cohen–Macaulay, Hilbert domain. Notice that $R = K[X, Y] \cap K(X^q, XY, Y^q)$.

(b) R is not factorial. The ideal $P = (X^q, XY)$ is prime, of height 1, and not principal. $\mathrm{Cl}(R)$ is cyclic of order q, generated by $\mathrm{div}(P)$. Notice that $P^q = X^q I$, where $(X^q, XY, Y^q)^{q-1} \subseteq I$; and $\mathrm{div}(P^q) = (X^q)$.

(c) Instead of a field, we can allow K to be any factorial domain of characteristic p.

(d) This example is from [Waterhouse 1971]. See also [Fossum 1973, p. 103].

□ □ □ □ □

Example 57

Let K_0 be a field of characteristic 2, let $L = K_0(t_1, t_2, \ldots)$, and let $K = K_0(t_1^2, t_2^2, \ldots)$. Let $R_0 = K[X, Y]$. Let $R = L[[X, Y]]$. For each $i > 0$, let $d_i = t_{2i}X + t_{2i-1}Y$. Let $T = R_0[d_1, d_2, \ldots]$ and let $L_0 = K(d_1, d_2, \ldots)((X, Y))$.

(a) R_0 and R are 2-dimensional, regular, local domains.

(b) R is integral over T and T is integral over R_0. Therefore T is quasi-local and 2-dimensional.

(c) T is not Noetherian, since the ideal $(t_1, t_2, \ldots)$ is not finitely generated.

(d) T is a Krull domain, since $T = R \cap L_0 =$ the intersection of a Krull domain with a subfield of its quotient field.

(e) Thus T is a 2-dimensional, non-Noetherian, quasi-local Krull domain, not the integral closure of a Noetherian domain. (The integral closure of a 2-dimensional Noetherian domain must be Noetherian.)

(f) This example is from [Eakin and Heinzer 1970].

□ □ □ □ □

Example 58

Let $R_0 = \mathbf{Q}[X, Y]$. Let R be the subring of R_0 consisting of all elements whose constant terms lie in $\mathbf{Z}$. (So $R = \mathbf{Z} + XR_0 + YR_0$.)

(a) R is 3-dimensional and integrally closed.

(b) R is not regularly integrally closed. To see this, consider the ideal $I = (X, Y)$. I is finitely generated, so $\mathrm{div}(I)$ is a finitely generated divisor. If $J = \mathbf{Q}[X, Y]$, J is a fractional ideal of R and $IJ = I$, so $\mathrm{div}(I) + \mathrm{div}(J) = \mathrm{div}(IJ) = \mathrm{div}(I)$, but $\mathrm{div}(J)$ is not the identity element $\mathrm{div}(R)$ in the monoid $D(R)$. Therefore, $\mathrm{div}(I)$ is not a regular element of $D(R)$, so R is not regularly integrally closed.

(c) Of course, R is not Noetherian.

(d) For each $n > 0$, let $R_n = \mathbf{Z}[X/n!, Y/n!]$. Then each R_n is factorial, and $R_1 \subseteq R_2 \subseteq R_3 \subseteq \ldots$.

(e) R is the union of the R_n. Thus, the union of an ascending chain of factorial domains need not be even regularly integrally closed.

(f) This example is from [Bourbaki 1972, pp. 554–555, Exercise 30, and pp. 561–562, Exercise 19].

Example 59

Let $p_1, p_2, \ldots$ be the sequence of all prime positive integers, and let K be the field containing $\mathbf{Q}$ and generated over $\mathbf{Q}$ by all p_ith roots of 1, for all $i > 0$. Let R be the integral closure of $\mathbf{Z}$ in K.

(a) R is 1-dimensional and integrally closed.

(b) Let K_n be the subfield of K containing $\mathbf{Q}$ and generated over $\mathbf{Q}$ by all p_ith roots of 1 for $1 \leqslant i \leqslant n$. Let R_n be the integral closure of $\mathbf{Z}$ in K_n. Then each R_n is a Dedekind domain, and $R_1 \subseteq R_2 \subseteq \ldots$, and R is the union of this chain.

(c) R is Prüfer, since it is the integral closure of a Prüfer domain in an algebraic extension of its quotient field.

(d) In fact, R is an almost Dedekind domain, i.e., R_P is a Noetherian valuation domain for each maximal ideal P of R. This is a rather deep fact.

(e) Each principal prime element of $\mathbf{Z}$ is contained in infinitely many prime ideals of R. Thus R is not Noetherian; so it is an almost Dedekind domain which is not Dedekind. One proof involves study of the relationship between prime ideals of R_n and prime ideals of R.

(f) This example is from [Nakano 1953].

Example 60

Let **N** *be the set of all positive integers. Let G be the set of all functions from* **N** *to* **Z** *such that f(n) is constant for all n greater than some integer, i.e., f takes on only finitely many values. G is an abelian group under pointwise addition, and G is lattice-ordered under the order f ⩾ 0 if and only if f(n) ⩾ 0 for all n > 0. Let R be a domain having G as its group of divisibility.*

(a) One such domain can be obtained by the method outlined in Chapter 2, Section B; this method actually gives us a Bézout domain with G as its group of divisibility.

(b) Since G is lattice-ordered, R is certainly pseudo-Bézout.

(c) R is almost Dedekind, that is, for each nonzero prime ideal P of R, R_P is a Noetherian valuation domain. To show this, we need to determine all such prime ideals P. The nonzero prime ideals of R correspond one-to-one to the proper convex sublattice subgroups of G. These subgroups are as follows: For $i > 0$, let H_i = the subgroup generated by all $f \in G$ such that $f \geqslant 0$ and $f(i) = 0$. Let H_0 = the subgroup generated by all $f \in G$ such that $f \geqslant 0$ and the eventual value of f is 0. These are all of the proper convex sublattice subgroups.

For $i \geqslant 0$, let P_i be the prime ideal of R corresponding to H_i. The group of divisibility of R_{P_i} is G/H_i (up to isomorphism), and for each $i \geqslant 0$, G/H_i is isomorphic to **Z**. Therefore each R_{P_i} is a Noetherian valuation domain.

(c) R is not Noetherian, so it is not a Dedekind domain. (One way to show R is not Noetherian is to show it does not satisfy the ascending chain condition on principal ideals, by producing an infinite descending chain of elements of G.)

(d) The polynomial ring $R[X]$ is an almost Krull domain that is not Krull. $R[X]$ is also pseudo-Bézout.

(e) This example is from [Heinzer and Ohm 1971].

□ □ □ □ □

Example 61

Let R be any ring. Let T = R[X, Y/X].

(a) T lies between $R[X, Y]$ and its total quotient ring, but T is not a localization of $R[X, Y]$. This is in contrast to QR domains.

(b) This example is from [Gilmer 1972a, p. 22].

□ □ □ □ □

Example 62

Let E, F_1, F_2, and G be fields, such that $E \subset F_1 \subset G$, $E \subset F_2 \subset G$, and $E = F_1 \cap F_2$. (So $F_1 \neq F_2$.) Let $R = E + X \cdot G[[X]]$, and let $V_i = F_i + X \cdot G[[X]]$, for $i = 1, 2$.

(a) R, V_1, and V_2 are each quasi-local and 1-dimensional. Also $R \subset V_1$, V_2 and $R = V_1 \cap V_2$.

(b) Although V_1 and V_2 lie between R and its quotient field, V_1 and V_2 are not localizations of R.

(c) We can arrange for R, V_1, and V_2 to be integrally closed, by requiring E to be algebraically closed in F_1 and F_2 and requiring F_1 and F_2 to be algebraically closed in G.

(d) This example is from [Kaplansky 1974b, pp. 78–79].

Example 63

Let m, n be nonnegative integers. Let $R_1 = \mathbf{Z}/(m)$ and let $R_2 = \mathbf{Z}/(n)$. Let $R = R_1 \oplus R_2$.

(a) R_1 and R_2 are both 0-dimensional, Noetherian, Hilbert rings. If m (or n) is prime, then R_1 (or R_2) is a domain. If m (or n) is square-free, then R_1 (or R_2) has no nilpotent elements and is von Neumann regular.

(b) R is also a 0-dimensional, Noetherian, Hilbert ring, and definitely not a domain, provided $m \neq 1 \neq n$.

(c) Let S be the set of all elements $(a, 0)$ of R such that a is a unit in R_1. Let $T = R_S$. T is isomorphic to R_1. If m is a prime integer, T may be a domain even though R is not. If m and n are relatively prime, R and T will be of different characteristics.

Example 64

Let $R = \mathbf{Z}/(16)$ and let $T = R[X]$.

(a) R is a 0-dimensional local ring with a nilpotent maximal ideal, generated by $\overline{2}$, the image of 2.

(b) T is a 1-dimensional Noetherian ring with a unique prime ideal, namely $\mathrm{rad}(T) = (\overline{2})$.

(c) The polynomial X^2 is in T. It is of degree 2, but it has 4 distinct roots in R, namely the images of 0, 4, 8, and 12. Therefore X^2 has several factorizations in T: (let $\overline{n}$ denote the image of n in R) $X^2 = X \cdot X = (X - \overline{4})(X - \overline{12}) = (X - \overline{8})^2$.

(d) Notice that this sort of behavior cannot occur in a domain.

(e) This example is from [van der Waerden 1970, vol. 1, p. 85].

Example 65

Let p be a prime integer and let $K = \mathbf{Z}/(p)$. Let $R = K[X^p, X^{p+1}]$, and let $T = R[Y - XY^p, Y^p]$.

(a) R is a 1-dimensional, Noetherian, Hilbert domain. Its integral closure is $K[X]$.

(b) T is 1 2-dimensional, Noetherian, Hilbert domain. Its integral closure is $K[X, Y]$.

(c) The rings $R[U, V]$ and $T[W]$ are isomorphic. To see this, consider the mapping $h: R[U, V] \rightarrow T[W]$ defined by mapping R identically onto itself, mapping U to $U' = Y - X(Y + XW)^p$, and mapping V to $V' = W + (Y + XW)^p$. Notice that $W = V' - (U' + XV')^p$, so the mapping is onto. In fact, this mapping h is an isomorphism.

(d) In spite of (c), T is not isomorphic to $R[U]$. This is a delicate nonisomorphism; for instance, $(R[U])^p$ is isomorphic to $T^p = R^p[Y^p]$, and $R[U][X^{-1}]$ is isomorphic to $T[X^{-1}] = K[X, X^{-1}, Y]$.

(e) This example is from [Hamann 1975a].

(f) It has been shown that it is possible to have isomorphic power series rings with nonisomorphic rings of coefficients. See [Hamann 1975b].

Example 66

Let K be a field of characteristic 2. *Let* $R = K[X^2, X^3, Y]/(Y^2)$ *and let* $T = K[X^2, X^3 + YZ, Y]/(Y^2)$. *Let* x, y, z *be the images of X, Y, Z, respectively, in either ring.*

(a) R is a 1-dimensional, Noetherian, Hilbert ring, obviously not a domain.

(b) T is a 1-dimensional, Noetherian, Hilbert ring also; in fact, T is isomorphic to R.

(c) Notice that both R and T are subrings of $K[X, Y, Z]/(Y^2)$, but $R \neq T$. The element z is algebraically independent of T.

(d) Let $R_0 = R[x]$ and $T_0 = T[x]$. R_0 and T_0 are not isomorphic, since $\mathrm{rad}(R_0)$ is principal and $\mathrm{rad}(T_0)$ is not. However, $R_0[z] = T_0[z] = K[x, y, z]$. The element z is algebraic over T_0.

(e) This example is from [Eakin and Kubota 1972].

Example 67

Let K be the field of real numbers. Let $R_0 = K[X, Y, Z]/(X^2 + Y^2 + Z^2 - 1)$ *and let* x, y, z *be the images of X, Y, Z, respectively. Let* $R = R_0[P, Q]$, *and let* $T = R_0[U, V, W]/(xU + yV + zW)$. *Let* u, v, w *be the images of U, V, W, respectively in T.*

(a) R_0 is a 2-dimensional, Cohen–Macaulay, Hilbert domain.

(b) R and T are both 4-dimensional, Cohen–Macaulay, Hilbert domains.

(c) R and T are not isomorphic. (This is nontrivial.)

(d) However, $R[t]$ and $T[t]$ are isomorphic, by the mapping $h: R[t] \to T[t]$ which maps R_0 identically onto itself ans sends u to $u' = yP - zQ$, v to $v' = zt - xP$, w to $w' = xQ - yt$, and t to $t' = yQ + xt + zP$. This mapping is onto, since $P = yu' - xv' + zt'$, $Q = xw' - zu' + yt'$, and $t = zv' - yw' + xt'$.

(e) In fact, R_0 is factorial, as in Example 55. Therefore, R, $R[t]$, $T[t]$, and T are all factorial; it is not obvious at first glance that T is factorial. Thus we have a pair of 4-dimensional, factorial, Cohen–Macaulay, Hilbert domains, in characteristic 0, which are not isomorphic, but such that the polynomial rings over them are isomorphic.

(f) This example is from [Hochster 1972].

Example 68

Let p be a prime integer and let $K = \mathbf{Z}/(p)$. Let $L = K(X_1, X_2, \ldots)$. Let $R = L^p + Y \cdot L[[Y]]$ and let $T = L[[Y]]$.

(a) R is a 1-dimensional quasi-local domain.

(b) T is a Noetherian valuation domain.

(c) T is the integral closure of R. Thus, T is Noetherian and integral over R, which is not Noetherian. Notice that T is not finitely generated over R as an R-module; if it were so, the two rings would have to be either both or neither Noetherian.

□ □ □ □ □

Example 69

Let $R = \mathbf{Z}[X]$ and let $T = R/(2X, X^2 - X)$. Let x be the image of X.

(a) $T = \mathbf{Z}[x]$, where $x^2 = x$ and $2x = 0$; thus T is integral over $\mathbf{Z}$.

(b) T is therefore a 1-dimensional, Noetherian, Hilbert ring.

(c) Let $P = (2, X - 1)$ in R, and let $P' = (2, x - 1)$ in T, so P' is the image of P. P' is maximal in T, and $\mathrm{ht}(P') = 0$, while $\mathrm{ht}(P' \cap \mathbf{Z}) = 1$. If $Q = (2, X)$ in R, and $Q' = (2, x)$ is its image in T, then Q' is maximal in T, and $\mathrm{ht}(Q') = 1 = \mathrm{ht}(Q' \cap \mathbf{Z})$. Also, $P' \cap \mathbf{Z} = Q' \cap \mathbf{Z} = (2)$.

(d) Thus GD fails for the pair $\mathbf{Z}$, T. This shows that in the going-down theorem, it is not sufficient to have a ring (with zero-divisors) integral over an integrally closed domain; both rings must be domains.

(e) Also notice that we have two prime ideals P' and Q', both lying over the same prime ideal (2) in $\mathbf{Z}$, but of different heights.

(f) This example is from [Gilmer 1972a, p. 109, Exercise 7].

□ □ □ □ □

Example 70

Let K be a field and let $T = K[X]$. Let $x_1 = X(X - 1)$ and $x_2 = X^2(X - 1)$. Let $R = K[x_1, x_2]$.

(a) T is a principal ideal domain.

(b) R is a 1-dimensional Noetherian domain.

(c) T is the integral closure of R, so $T[Y]$ is the integral closure of $R[Y]$. However, GD fails for the extension $T[Y]$ of $R[Y]$. Let $Q = (X - 1, Y)$ in $T[Y]$, and let $P = (Yx_1 - x_2)$ in $R[Y]$. P and Q are both prime ideals. Q lies over the prime ideal $Q \cap R[Y] = (x_1, x_2, Y)$ in $R[Y]$, and $P \subseteq (x_1, x_2, Y)$; but there is no prime ideal of $T[Y]$ contained in Q and lying over P.

(d) The going-down theorem does not apply in this case, because R is not integrally closed.

(e) This example is from [Matsumura 1970, pp. 32–33].

Example 71

Let $T = K[X, Y, Z]/(X^2, XY, XZ, YZ - Y, Z^2 - Z)$, where K is a field. Let x, y, z be the images of X, Y, Z, respectively. Let $R = K[x, y]$.

(a) R and T are both 1-dimensional Noetherian rings; clearly not domains.

(b) R is integrally closed in its total quotient ring of fractions.

(c) T is integral over R. However, GD fails for the pair R, T. In R, let $P = (x)$ and $Q = (x, y)$. In T, let $Q' = (x, y, z)$. Then Q', Q, and P are all prime ideals, Q' lies over Q, $P \subseteq Q$, but there is no prime ideal P' of T, contained in Q' and lying over P.

(d) In this case, the going-down theorem does not apply because R is not a domain.

(e) This example is from [Bourbaki 1972, p. 366, Exercise 13].

Example 72

Let $R = K[X, Y, Z]/(Y^2 - X^2 - X^3)$, where K is a field of characteristic 0.

Let x, y, z be the images of X, Y, Z, respectively. Let $T = K[x, y/x, z]$.

(a) T is a 2-dimensional, integrally closed, Cohen–Macaulay, Hilbert domain.

(b) R is a 2-dimensional, Cohen–Macaulay, Hilbert domain.

(c) T is the integral closure of R. However, GD fails for the pair R, T. In R, let $P = (xz - y, z^2 - 1 - x)$ and let $Q = (x, y, z - 1)$. P and Q are prime ideals of R. In T, the maximal ideal $Q' = (x, 1 + y/x, z - 1)$ lies over Q; but there is no prime ideal P' of T, such that P' is contained in Q' and lies over P.

(d) Notice the similarity between this example and Example 70.

(e) This example is from [Bourbaki 1972, p. 365, Exercise 13].

Example 73

Let R be any integral domain containing a maximal ideal M of height $n = \dim(R) > 0$. Let $(0) = P_0 \subset P_1 \subset \ldots \subset P_n = M$ be a chain of prime ideals in R. For each integer k, $0 \leqslant k \leqslant n$, let $R_k = R/P_k$. Let $T = R_0 \oplus R_1 \oplus \ldots \oplus R_n$.

(a) T is clearly not a domain, and $\dim(T) = \dim(R)$.

(b) We can embed R in T by the map $R \to T$ defined by sending each $x \in R$ to $(x_0, x_1, \ldots, x_n)$, where x_k is the image of x in R_k, for $0 \leqslant k \leqslant n$.

(c) T is integral over R. To see this, notice that T is generated over R by the elements $e_k = (0, \ldots, 0, 1_k, 0, \ldots, 0)$, where 1_k is the identity element of R_k, and each e_k is idempotent.

(d) For $0 \leqslant k \leqslant n$, let $M_k = R_0 \oplus \ldots \oplus R_{n-k-1} \oplus (M/P_{n-k}) \oplus R_{n-k+1} \oplus \ldots \oplus R_n$. Then M_k is a maximal ideal of T, M_k lies over M, and $\mathrm{ht}(M_k) = k$. Thus GD fails for R, T.

(e) By omitting some of the summands R_k, we could obtain an integral extension of R having maximal ideals lying over M of precisely the heights we want.

(f) This example is an expanded version of [Kaplansky 1974b, p. 43, Exercise 25].

Example 74

Let $R = K[X, Y]/(X^3 - X^2 + Y^2)$, where K is an algebraically closed field. Let x, y be the images of X, Y, respectively. Let $T = K[x, y/x]$.

(a) T is a principal ideal domain, since actually $T = K[y/x]$. (Notice that $1 - (y/x)^2 = x$.)

(b) R is a 1-dimensional Noetherian domain.

(c) T is the integral closure of R.

(d) Let $P = (x, y)$ in R. Then P is a maximal ideal in R, and P^2 is P-primary. Also, $P^2 T \cap R = P^2$. However, no primary ideal of T lies over P^2 in R. Thus LO fails for primary ideals.

(e) Notice the similarity between this example and Example 72.

(f) This example is from [Gilmer 1972a, p. 113, Exercise 27].

Example 75

Let R be the integral closure of $\mathbf{Z}$ in $\mathbf{Q}(\sqrt{d})$, where $d \in \mathbf{Z}$ is square-free, i.e., not divisible by a square.

(a) R is a Dedekind domain.

(b) If $d - 1$ is not divisible by 4, then $R = \mathbf{Z}[\sqrt{d}]$.

(c) If $d - 1$ is divisible by 4, then $R = \mathbf{Z}[t]$, where $t = (1 + \sqrt{d})/2$.

(d) Notice that R is in all cases a *simple* integral extension of $\mathbf{Z}$, i.e., generated over $\mathbf{Z}$ by one additional element.

Example 76

Let R be the integral closure of $\mathbf{Z}$ in $\mathbf{Q}(t)$, where $t^3 = m$ and $m \in \mathbf{Z}$ is square-free.

(a) R is a Dedekind domain.

(b) If neither $m - 1$ nor $m + 1$ is divisible by 9, then $R = \mathbf{Z}[t]$.

(c) If $m - 1$ is divisible by 9, then $R = \mathbf{Z}[s, t]$, where $s = (1 + t + t^2)/3$.

(d) If $m + 1$ is divisible by 9, then $R = \mathbf{Z}[u, t]$, where $u = (1 - t + t^2)/3$.

(e) Notice that in cases (c) and (d), R is not a simple integral extension of $\mathbf{Z}$. This is in contrast to the situation with fields; any finite-dimensional extension of $\mathbf{Q}$ is simple.

(f) This example does not cover all cube roots of integers. For instance, if $t^3 = 4$, this is not covered since 4 is divisible by a square. However, if $t^3 = 4$, then $\mathbf{Q}(t) = \mathbf{Q}(t')$, where $(t')^3 = 2$. Here $t' = t^2/2$.

Example 77

Let $K = \mathbf{Z}/(p)$, where p is a prime integer. Let $L = K(X_1, X_2, \ldots)$. Notice that $L^p = K(X_1^p, X_2^p, \ldots)$. Let G be the set of all fields L_a such that $L^p \subseteq L_a \subseteq L$ and L_a is a finite algebraic extension of L^p. For each such field L_a, let $R_a = L_a[[Y]]$. Let $T = L[[Y]]$.

(a) Each ring R_a is a Noetherian valuation domain, with maximal ideal $M_a = (Y)$.

(b) T is also a Noetherian valuation domain, with maximal ideal $N = (Y)$.

(c) Let R be the union of the domains R_a. Then R is a ring. (To add or multiply s, $t \in R$, we have $s \in R_j$, $t \in R_k$; for some c, R_c contains R_j and R_k, so the sum or product of s and t lies in $R_c \subseteq R$.) Notice that $R \neq T$, since, for instance, the element $r = X_1 Y + X_2 Y^2 + \ldots$ is in T but not in R.

(d) In fact, R is a Noetherian valuation domain, with maximal ideal $M = (Y)$ and residue field $F = $ union of the fields L_a. To see this, observe that $T^p \subseteq R_a \subseteq R \subseteq T$, and T is obviously integral over T^p, so T is integral over R and over each ring R_a. Thus R is quasi-local and 1-dimensional, by inheritance from T. The maximal ideal M is the union of the maximal ideals M_a of the rings R_a; each $M_a = Y \cdot R_a$; so $M = Y \cdot (\text{the union of the rings } R_a) = Y \cdot R = (Y)$. The other parts of the statement are similarly easy to see.

(e) Let L_1 be the quotient field of R, and let R_1 be the integral closure of R in $L_1(r)$, for r as in (c). Notice that $r \in R_1$, so the quotient field of R_1 is $L_1(r)$. $L_1(r)$ is a finite algebraic extension of L_1.

(f) In fact, R_1 is a Noetherian valuation domain, with maximal ideal (Y). Why? Since T is integral over R and $R \subseteq R_1 \subseteq T$, T is also integral over R_1. Therefore, R_1 is 1-dimensional and quasi-local. It is integrally closed by definition. Suppose M_1 is the maximal ideal of R_1. Then if $g \in M_1$, $g \in N$, so $g = Yf$

for some $f \in T$. Then $f = g/Y \in L_1(r)$. Since $f \in T$, f is integral over R_1, so $f \in R_1$ since R_1 is integrally closed. Hence $M_1 = Y \cdot R_1$ and R_1 is actually Noetherian.

(g) R_1 is not a finitely generated R-module. To see this, notice that $R_1 = R + Y \cdot R_1$, and Y is in the Jacobson radical of R. If R_1 were a finitely generated R-module, Nakayama's lemma would imply that $R = R_1$, which is a contradiction.

(h) The whole point of this example is in (g); that R_1 is the integral closure of R, a domain, in a finite algebraic extension of the quotient field of R, and yet R_1 is not a finitely generated R-module. What makes this possible is the fact that $L_1(t)$ is not a separable extension of L_1. It is also singularly nice that all of the domains involved are Noetherian valuation domains, which are rather well-behaved rings, in most ways.

(i) This example is from [Murthy 1976, vol. 1, pp. 88–90].

Example 78

Let $T = K[[X]]$, where K is a field of characteristic 2. Let $u \in T$, and let $L_1 = K(X, u)$ and let $L_2 = K(X, u^2)$. Let $R_1 = T \cap L_1$ and $R_2 = T \cap L_2$.

(a) T, R_1, and R_2 are all Noetherian valuation domains. For T, this is obvious; for R_1, notice R_1 is the intersection of a Krull domain with a subfield of its quotient field, so R_1 is a Krull domain. Its group of divisibility is a subgroup of the value group of T, and is therefore infinite cyclic; thus R_1 is a Noetherian valuation domain. The same remarks hold for R_2.

(b) R_1 is the integral closure of R_2 in L_1.

(c) Either $L_1 = L_2$ or L_1 is 2-dimensional as an algebra over L_2. If we choose $u = X$, then $L_1 = L_2$. If u is transcendental over $K(X)$, then $u \notin L_2$, and L_1 is definitely 2-dimensional over L_2.

(d) If L_1 is 2-dimensional over L_2, then R_1 is not a finitely generated R_2-module.

(e) It is actually possible to pick u transcendental over $K(X)$; see Example 102.

(f) This example is from [Kaplansky 1974b, Theorem 100].

□ □ □ □ □

Example 79

Let $K_0 = \mathbf{Z}/(p)$, where p is a prime integer. Let $T = L[[Y, W]]$ and let $R_0 = L^p[[Y, W]] \otimes_{L^p} L$.

(a) T is a 2-dimensional, complete, regular, local domain.

(b) R_0 is a 2-dimensional local domain whose completion is T. T is integral over R_0.

(c) For $n > 0$, let $c_n = X_n Y + X_{n+1} YW + X_{n+2} YW^2 + \ldots$. Each $c_n \in T$. Let $R_1 = R_0[c_0]$ and let $R_2 = R_0[c_0, c_1, c_2, \ldots]$.

(d) R_1 and R_2 are both integral over R_0, and T is integral over each of them. Therefore R_1 and R_2 are both 2-dimensional and quasi-local.

(e) R_1 is Noetherian, since it is finitely generated over R_0, which is Noetherian. R_2 is not Noetherian, since $(c_0) \subset (c_0, c_1) \subset (c_0, c_1, c_2) \subset \ldots$ is an infinite properly ascending chain of ideals.

(f) R_1 is not integrally closed, since, for instance, c_0/Y is in the quotient field of R_1, not in R_1, and $(c_0/Y)^p$ is in R_1. R_2 is integral over R_1, clearly, since it is integral over $R_0 \subseteq R_1$. R_2 is contained in the quotient field of R_1 since $c_n W^n - c_0$ is an element of $R_0 \subseteq R_1$.

(g) Thus R_2, which is not Noetherian, lies between R_1, which is Noetherian, and its integral closure, which must be Noetherian. This is in contrast to the behavior in the 1-dimensional case.

(h) This example is from [Bourbaki 1972, p. 359], and is due to Nagata ([Nagata 1962, p. 207]).

□ □ □ □ □

Example 80

Let $K_0 = \mathbf{Z}/(2)$. Let $L = K_0(X_0, X_1, \ldots)$. Let $T = L[[U, V, W]]$ and let $R_0 = L^p[[U, V, W]] \otimes_{L^p} L$.

(a) T is a 3-dimensional, complete, regular, local domain.

(b) T is integral over R_0, so R_0 is 3-dimensional and quasi-local. It can be shown that R_0 is actually Noetherian, and that T is the completion of R_0. Notice that if $J = (U, V, W)$, then R_0 contains $T_0 = L[U, V, W]_J$, with the maximal ideal of R_0 lying over the maximal ideal of T_0; the completion of T_0 is T also.

(c) Let $b = U(X_0 + X_2 W + X_4 W^2 + \ldots) + V(X_1 + X_3 W + X_5 W^2 + \ldots)$.

Let $R = R_0[b]$. R is Noetherian, since it is finitely generated over R_0, which is Noetherian, and R is 3-dimensional and quasi-local, since T is integral over R.

(d) Let $S =$ the integral closure of R. It is easy to see that S is 3-dimensional and quasi-local. In fact, $S = R_0[b_0, b_1, b_2, \ldots]$, where $b_0 = b$ and for $n > 0$, $b_n = U(X_{2n} + X_{2n+2}W + X_{2n+4}W^2 + \ldots) + V(X_{2n+1} + X_{2n+3}W + X_{2n+5}W^2 + \ldots)$. Since $(b_0) \subset (b_0, b_1) \subset (b_0, b_1, b_2) \subset \ldots$ is an infinite properly ascending chain of ideals, S is not Noetherian.

(e) Thus R is a 3-dimensional local domain, of characteristic 2, whose integral closure is quasi-local but not Noetherian.

(f) It has been proven that in this example S is actually a factorial domain. Thus S is 3-dimensional, quasi-local, and factorial, but not Noetherian.

(g) This example is from [Bourbaki 1972, pp. 359–360], and is due to Nagata [Nagata 1962, p. 208]. See [David 1972] for the proof that S is factorial. (I do not know whether the completion of S is factorial.)

Example 81

Let $T = K[X, Y]$, where K is a field and let $R = K[X^2, Y^2, XY, X^3, Y^3, XY^2, X^2Y]$.

(a) T is a 2-dimensional, regular, factorial, Hilbert domain.

(b) R is a 2-dimensional, Noetherian, Hilbert domain.

(c) T is the integral closure of R.

(d) Even though T is regular and is integral over R, R is not even Cohen–Macaulay. Specifically, let $M = (X, Y) \cap R$; M is a maximal ideal in R. Consider R_M. We have $X^4 \notin (X^3)$ in R, but $X^4 M \subseteq (X^3)$. Therefore, if X^3 is the first term of an R-sequence, there can be no second term from M. Thus R_M is not Cohen–Macaulay, and neither is R.

(e) This example is from [Matsumura 1970, p. 109].

Example 82

Let $K = \mathbf{Q}(\sqrt{-5})$ *and let* $L = \mathbf{Q}(\sqrt{-5}, \sqrt{-1})$. *Let* R *be the integral closure of* **Z** *in* K. *Let* T *be the integral closure of* **Z** *in* L.

(a) R is a Dedekind domain, but not a principal ideal domain; e.g., the ideal $(3, 2 + \sqrt{-5})$ is not principal.

(b) T is a principal ideal domain.

(c) Notice that T is integral over R, and T is factorial, but R is not factorial; and R is integral over **Z**, which is factorial.

(d) This example is from [Gilmer 1972a, p. 505, Exercise 3]; also see [Cohn 1962, p. 236].

Example 83

Let $R = K[X^2 - 1]$, *where* K *is a field and let* $T = K[X]$.

(a) R and T are both principal ideal domains.

(b) T is integral over R. Let $M = (X^2 - 1)$ in R and let $N = (X - 1)$ in T. These are both maximal ideals and M lies over N.

(c) T_N is not integral over R_M. The integral closure of R_M in T_N is T_S, where S is the complement of M in R. Thus T_N is a localization of T_S, so there are elements of T_N which are not integral over R_M, such as $(X + 1)^{-1}$

(d) This example is from [Bourbaki 1972, p. 314], and from [Gilmer 1972a, p. 132, Exercise 14].

Example 84

Let $K = \mathbf{Q}(\sqrt{-5})$. *Let* R *be the integral closure of* **Z** *in* K.

(a) R is a Dedekind domain and integral over **Z**.

(b) Let $M = (3, 2 + \sqrt{-5})$ in R. M is a maximal ideal in R and lies over the ideal $J = (3)$ in **Z**.

(c) R_M is not integral over $\mathbf{Z}_J$, since there are elements of R_M which are not integral over $\mathbf{Z}_J$, such as $(1 + \sqrt{-5})^{-1}$.

(d) Notice the similarities between this example and Example 83.

Example 85

Let R be a ring containing nonzero elements u, v such that $(u) \cap (v) = (0)$. Let $T = R[X, Y]/(uX, vY)$. Let x, y be the images of X, Y, respectively.

(a) One way to obtain such a ring R is to let $R = K[U, V]/I$, where K is a field and $I = (U) \cap (V)$.

(b) R is a subring of T. The elements x and y are algebraic over R, since $ux = vy = 0$; but the element $x + y$ is not algebraic over R.

(c) It is because of the phenomenon in (b), i.e., that the sum of elements each algebraic over a ring may not be algebraic, that ring theorists deal with integral extensions rather than the more general algebraic extensions.

(d) This example is from [Gilmer 1969b].

Example 86

Let $T = K[X, Y]/(X^2, XY)$, where K is a field. Let x, y be the images of X, Y, respectively. Let $R = K[x]$.

(a) R is a 0-dimensional local ring.

(b) T is a 1-dimensional Noetherian ring. Thus T is not integral over R; R is integrally closed in T.

(c) No element of T is algebraically free over R. This sort of behavior—that T is an algebraic extension of R, but R is integrally closed in T—cannot occur in domains. (If T and R were domains and $t \in T$ is algebraic over R, so $a_0 t^n + a_1 t^{n-1} + \ldots + a_n = 0$, with all $a_i \in R$, then $a_0 t$ is integral over R. In this case, we have $xy = 0$, $n = 1$.)

(d) This example is from [Bourbaki 1972, p. 370, Exercise 2].

Example 87

Let $R = K[X, YX^2, Y^2X^3, Y^3X^4, \ldots]$, where K is a field.

(a) R is a 2-dimensional integrally closed domain. To see these facts, notice that R is isomorphic to the ring of Example 3, since $R = K[X, X(XY), X(XY)^2, X(XY)^3, \ldots]$.

(b) R is not completely integrally closed, since XY is in the quotient field $K(X, Y)$ of R, and is almost integral over R, but $XY \notin R$. That is, $R[XY]$ is contained in the finitely generated R-module $X^{-1}R$, but is not itself a finitely generated R-module; thus XY is almost integral, but not integral, over R.

(c) Of course, R is not Noetherian.

(d) This example (or Example 3) is about as simple and explicit as one can find, to see that a submodule of a finitely generated module need not be finitely generated.

(e) This example is from [Bourbaki 1972, p. 355, Exercise 2].

Example 88

Let $L = K(X_1, X_2, \ldots)$, where K is a field. For $n > 0$, let $L_n = K(X_1, X_2, \ldots, X_n)$. Let $R = K + Y \cdot L[[Y]]$. Let $T = R[X_1, X_2, \ldots]$. For $n > 0$, let $R_n = K + Y \cdot L_n[[Y]]$, and let $T_n = R_n[X_1, \ldots, X_n]$.

(a) R and each of the rings R_n are 1-dimensional, quasi-local, non-Noetherian domains.

(b) T is almost integral over R, and for each $n > 0$, T_n is almost integral over R_n. However, $\dim(T_n) = n + 1$ and T is infinite-dimensional. Thus $\dim(R) < \dim(T)$ and $\dim(R_n) < \dim(T_n)$, in contrast to integral extensions.

Example 89

Let V be a valuation domain with no prime ideals of height 1. (For instance, see Example 35.) Let K be the quotient field of V. Let $V_0 = V$, and for $n > 0$, let $V_n = V[X_1, \ldots, X_n]$ and let R_n be the complete integral closure of V_n.

(a) For $n > 0$, V_n is an infinite-dimensional integrally closed domain.

(b) $R_0 = K$ (as in Example 35), so for $n > 0$, R_n is just $K[X_1, X_2, \ldots, X_n]$, which is an n-dimensional, regular, factorial domain.

(c) For $n > 0$, R_n is almost integral over V_n. However, $\dim(R_n) = n$.

(d) This example and Example 88, together, indicate that in an almost integral ring extension, the dimensions of the rings are not related in any obvious fashion.

Example 90

Let $R = \mathbf{Z} + X \cdot \mathbf{Q}[X]$. *Let* $T = \mathbf{Q}[X]$.

(a) T is a principal ideal domain.

(b) R is a 2-dimensional integrally closed domain.

(c) The complete integral closure of R is T. Notice the similarity between this example and Example 88.

(d) This example is from [Gilmer 1972a, p. 144, Exercise 2].

Example 91

Let $T = K[X, Y]$, *where* K *is a field. Let* $S = K[X, XY, XY^2, \ldots]$. *Let* $R = K[X, X^3 Y^3, \ldots, X^{2n+1} Y^{n(2n+1)}, \ldots]$.

(a) T is a 2-dimensional, regular, factorial domain.

(b) S is the ring of Example 3; 2-dimensional, integrally closed, and not completely integrally closed.

(c) R is a 2-dimensional domain, but not integrally closed and not Noetherian.

(d) R, S, and T all have the same quotient field, namely $K(X, Y)$.

(e) The complete integral closure of R is S, and the complete integral closure of S is T. Thus R is a domain whose complete integral closure is not completely integrally closed.

(f) This example is from [Gilmer and Heinzer 1966].

Example 92

Let $H = \mathbf{Z} \oplus \mathbf{Z}$, with the lexicographic order, i.e., $(a, b) > 0$ if $a > 0$ or if $a = 0$ and $b > 0$. Let $\mathbf{Z}^+$ be the set of nonnegative integers. Let G be the set of all functions from $\mathbf{Z}^+$ to H having finite support, i.e., if $f \in G$, then $f(n) = 0$ for all but finitely many $n \in \mathbf{Z}^+$. G is an abelian group with addition defined pointwise, and G is lattice-ordered by defining $f \geqslant 0$ if $f(n) \geqslant 0$ for all $n \in \mathbf{Z}^+$. Let R be a domain having G as its group of divisibility. Let K be the quotient field of R.

(a) In fact, we can require R to be a Bézout domain, by using the construction given in Chapter 2, Section B.

(b) Each nonzero element x of K corresponds to a function x' from $\mathbf{Z}^+$ to H, where $x'(n) = (f/g)(n) = f(n) - g(n)$, for some $f, g \in R$.

(c) R is infinite-dimensional.

(d) Let S be the complete integral closure of R in K. The nonzero elements of S consist of all the $x \in K$ which are not 0 and such that $x'(n) = (a_n, b_n)$ with $a_n \geqslant 0$ for all n and $b_n \geqslant 0$ for all but finitely many n.

(e) S is not completely integrally closed, since if $y \in K$ has $y'(n) = (0, -1)$ for all n, then y is almost integral over S, but is not in S.

(f) Let T be the complete integral closure of S in K. The nonzero elements of T consist of all the $x \in K$ which are not 0 and such that $x'(n) = (a_n, b_n)$ with $a_n \geqslant 0$ for all n.

(g) For $n \geqslant 0$, let $V_n = $ the subring of K consisting of all $x \in K$ such that $x = 0$ or $x'(n) \geqslant (0,0)$ in H. Let W_n be the subring of K consisting of all $x \in K$ such that $x = 0$ or $x'(n) = (a_n, b_n)$ with $a_n \geqslant 0$.

(h) For each $n \geqslant 0$, V_n is a 2-dimensional valuation domain, and W_n is the unique 1-dimensional valuation domain containing V_n and contained in K. R is the intersection of all the V_n, and T is the intersection of all the W_n. Therefore T is completely integrally closed.

(i) This example is from [Heinzer 1969]. See also item #5 in the Further References section.

□ □ □ □ □

Example 93

Let R be a domain which is integrally closed, but not completely integrally closed. Let K be the quotient field of R. Let a be a nonzero noninvertible ele-

ment of R such that there is a nonzero element d of R with $da^{-n} \in R$ for all $n > 0$. (For instance, R might be a 2-dimensional valuation domain, and a might then be any element of the maximal ideal which is not in the (unique) prime ideal of height 1.) Let $T = R[[X]]$.

(a) There is, in the quotient field of T, an element f satisfying $f^2 - af + X = 0$, but which is not in $R[[X]]$. Thus $R[[X]]$ is not integrally closed. (Specifically, if $\text{char}(K) \neq 2$, we have $f = d^{-1}(da - dX/a + 2dX^2/a^3 - \ldots)$. Notice that every coefficient is in R. If $\text{char}(K) = 2$, we have $f = d^{-1}(da + dX/a + dX^2/a^3 + dX^4/a^7 + \ldots)$. Again, every coefficient is in R.)

(b) This example demonstrates that if R is a domain which is not completely integrally closed, then $R[[X]]$ is definitely not integrally closed.

(c) This example is from [Bourbaki 1972, p. 361, Exercise 27].

Example 94

Let $R = K(U)[[X, Y, Z]]$, where K is a field. Let $f = X^2 + Y^3 + UZ^6$. Let $T = R/(f)$.

(a) T is a 2-dimensional, factorial, complete, local domain.

(b) $T[[V]]$ is not factorial. This (a factorial domain having a nonfactorial power series ring) is in contrast to the corresponding situation involving polynomials.

(c) Let $T_0 = K(U)[X, Y, Z, V]/(f)$. Let x, y, z, v be the images of X, Y, Z, V, respectively. Let $M = (x, y, z, v)$. T is the completion of the local domain $(T_0)_M$ in the M-adic topology. Thus $(T_0)_M$ is a local factorial domain whose completion is not factorial.

(d) Suppose i and j are both odd positive integers, not both equal to 3; then the domain $S = K(U)[[X, Y, Z]]/(g)$, where $g = X^2 + Y^i + UZ^{2j}$, is factorial, but $S[[V]]$ is not factorial.

(e) This rather famous example is from [Fossum 1973, pp. 116–117], and is originally due to Salmon ([Salmon 1966]).

Example 95

Let $R = Q[X, Y, Z]/(2X^3 + 4Y^3 - 3Z^3)$. Let x, y, z be the images of X, Y, Z, respectively. Let $M = (x, y, z)$ and let $T = R_M$.

(a) T is a 2-dimensional, factorial, local domain.

(b) If $\overline{T}$ is the completion of T in the M-adic topology, then $\overline{T}$ is also factorial.

(c) However, $T[[U]]$ is not factorial.

(d) Let $T_0 = K[x, y, z, U]$ and let $N = (x, y, z, U)$. Then $(T_0)_N$ is a local factorial domain whose completion (in the N-adic topology) is $\overline{T}[[U]]$, which is not factorial.

(e) In this example and in Example 94, the ring T is 2-dimensional. If S is a 1-dimensional factorial domain, then S must be regular, and the power series ring $S[[V]]$ must also be regular and factorial. On the other hand, if S is a complete, local, factorial domain which is of grade at least 3, then $S[[V]]$ is also factorial.

(f) In this example, there are many alternate choices available in place of the polynomial $2X^3 + 4Y^3 - 3Z^3$.

(g) This example is from [Fossum 1973, p. 117] and is originally due to Danilov ([Danilov 1970]).

□ □ □ □ □

Example 96

Let $R = Z[[X]]$. Let K be the quotient field of R.

(a) K is a proper subfield of $Q((X)) = Q[[X]][X^{-1}]$. K does contain $Q(X)$ and $R[1/n]$ for every n. Specifically, let $b = 1 + X/2 + X^2/3 + X^3/4 + X^4/5 + \dots$. Then $b \in Q((X))$, but $b \notin K$.

(b) In fact, there are elements of $Q((X))$ which are integral over R, but which are not in K. Since R is integrally closed, this also shows $K \neq Q((X))$. Specifically, let $t = 1 - X/2 - X^2/8 - X^3/16 - X^4/32 - \dots$; then $t^2 = 1 - X$, so t is integral over R.

(c) In fact, if we let $T = Z_{(2)}[[X]]$, then t is integral over T, is in $Q((X))$, and is not in the quotient field of T.

(d) It would be interesting to know what the integral closure of R in $Q((X))$ looks like.

□ □ □ □ □

Example 97

Let A be a well-ordered set whose ordinal type is that of $[0, \Omega)$. (That is, up to, but not including, the first uncountable ordinal.) For each $a \in A$, let $G_a = \mathbf{Z}$. Let G be the direct sum of the groups G_a; a typical element is a function from A to $\mathbf{Z}$ *having finite support, i.e., $f(a) = 0$ for all but finitely many $a \in A$. Define a linear order on G by declaring $f > 0$ if and only if there is $a \in A$ with $f(a) > 0$ and $f(b) = 0$ for $b > a$. (This is reverse lexicographic order.) G is a linearly ordered abelian group. Let V be a valuation domain whose value group is isomorphic to G.*

(a) V is infinite-dimensional.

(b) The intersection of any sequence of nonzero ideals of V is nonzero. It is enough to prove this for a sequence of principal ideals. If t_1, t_2, t_3, ... are nonzero elements of V, these correspond to g_1, g_2, g_3, ..., nonzero positive elements of G. Since each g_i has finite support as a function from A to $\mathbf{Z}$, and since the set A is sufficiently large, there is an $a \in A$ with $g_i(a) > 0$ for all $i > 0$. There is a positive element $h \in G$ with $h(a) > 0$; then $h > g_i$ for all $i > 0$. The element h corresponds to $s \in V$, and s is divisible by t_i for all $i > 0$. Therefore $s \in (t_1) \subseteq (t_2) \subseteq (t_3) \subseteq \ldots$.

(c) Let K be the quotient field of V. The quotient field of $V[[X]]$ is precisely $K((X))$. If $f \in K[[X]]$, then $f = (s_0/t_0) + (s_1/t_1)X + (s_2/t_2)X^2 + \ldots$, for suitable elements s_i, t_i of V. Let s be a nonzero element of V divisible by t_i for all $i > 0$. Then $f = sf/s$, where $sf \in V[[X]]$, so f is in the quotient field of $V[[X]]$.

(d) In general, if R is a domain with quotient field L, then the quotient field of $R[[X]]$ is much smaller than $L((X))$. For these two fields to be equal, it is necessary and sufficient for the intersection of any sequence of nonzero ideals of R to be nonzero. (That is, it must be possible to rationalize a sequence of denominators all at once.) This is a very strong, unusual condition; it fails in any Noetherian domain not a field.

(e) This example is from [Gilmer 1967].

□ □ □ □ □

Example 98

Let $K = \mathbf{Z}/(p)$, where p is a prime integer. Let $R = K[X_0, X_1, X_2, \ldots]/(X_0^p, X_1^p, X_2^p, \ldots)$. Let x_i be the image of X_i, for each i. Let $T = R[[Y]]$.

(a) Let $f = x_0 + x_1 Y + x_2 Y^2 + \ldots$. For $n > 0$, let $f_n = x_0 + x_1 Y + \ldots + x_n Y^n$. Notice that $f^p = 0$, and for $n > 0$, $f_n^p = 0$. However, the ideal $I = (x_0, x_1, x_2, \ldots)$, which is generated by the coefficients of f, is not nilpotent, since for $n > 0$, $x_0 x_1 \ldots x_n$ is not 0 and is in I^{n+1}.

(b) Although $f^p = 0$, there is no element $r \in T$ with $rf = 0$. However, for each $n > 0$, we can let $r_n = x_0^{n-1} x_1^{n-1} \ldots x_n^{n-1}$; then $r_n f_n = 0$.

(c) This is in contrast to the parallel situation involving polynomials. If S is a ring, and $g \in S[Y]$, g is nilpotent if and only if each of its coefficients is nilpotent, which in turn implies that the coefficients generate a nilpotent ideal.

(d) This example is from [Fields 1969].

Example 99

Let $R = \mathbf{Q}[X_0, X_1, X_2, \ldots]/(X_0^n, X_1^n, X_2^n, \ldots)$, *where* $n > 1$. *Let* x_i *be the image of* X_i, *for each i. Let* $T = R[[Y]]$.

(a) Let $f = x_0 + x_1 Y + x_2 Y^2 + \ldots$. For $k \geqslant 0$, let $f_k = x_0 + x_1 Y + \ldots + x_k Y^k$. For each k, x_k is nilpotent and f_k is nilpotent. However, f is not nilpotent.

(b) T is a complete, local, 1-dimensional ring in which $f_0, f_1, f_2, \ldots$ is a Cauchy sequence of nilpotent elements whose limit is f, which is not nilpotent.

(c) This example is from [Fields 1969].

Example 100

Let S *be any ring. Let* $R = S[Y, X_0, X_1, X_2, \ldots]/(X_0 Y, X_0 - X_1 Y, X_1 - X_2 Y, \ldots, X_i - X_{i+1} Y, \ldots)$. *Let* $y, x_0, x_1, x_2, \ldots$ *be the images of* $Y, X_0, X_1, X_2, \ldots$, *respectively. Let* $T = R[[Z]]$.

(a) Let $f = y - Z$ and $g = x_0 + x_1 Z + x_2 Z^2 + \ldots$. Then $fg = 0$. All of the coefficients of g are zero-divisors, but f has a unit coefficient.

(b) This is in contrast to the parallel situation involving polynomials. If S is a ring, and $h \in S[Y]$, h is a zero-divisor if and only if there is a nonzero element

$s \in S$ with $sh = 0$.

(c) This example is from [Fields 1969].

□ □ □ □ □

Example 101

Let $R = K[[X]]$, where K is a field.

(a) Let $u = a_1 X + a_2 X^2 + a_3 X^6 + \ldots + a_i X^{i!} + \ldots$, where infinitely many of the coefficients are not 0. Then u and X are algebraically independent over K. That is, there is no polynomial f with $f(X, u) = 0$; if there were, it could be shown that u could not have so many zero coefficients.

(b) This is similar to the usual examples of transcendental real numbers, such as $0.101001000100001\ldots$, which have increasingly large gaps between nonzero digits.

(c) We used the existence of elements such as u in Example 28.

(d) The transcendence degree of $K((X))$ over $K(X)$ is uncountably infinite; this is not hard to see, especially if K is countable (so $K(X)$ is countable and $K((X))$ is uncountable).

(e) This example is from [Gilmer 1972a, p. 507, Exercise 15].

□ □ □ □ □

Example 102

Let $R = K[X, Y]/(XY - Y)$, where K is a field. Let x, y be the images of X, Y, respectively.

(a) R is a 1-dimensional, Noetherian, Hilbert ring.

(b) Let $M = (x, y)$. Then $\cap M^n = (y)$, so the M-adic topology on R is not Hausdorff.

(c) In forming the completion $\overline{R}$ of R, two Cauchy sequences are equivalent if their limits cannot be separated by open sets; thus the inclusion of R into $\overline{R}$ is not one-to-one, with (y) as its kernel.

(d) Thus $\overline{R}$ is actually Hausdorff.

□ □ □ □ □

Example 103

Let $R = K[X, Y]/(Y^2 - X^3 - X^2)$, where K is a field with char$(K) \neq 2$. *Let x, y be the images of X, Y, respectively.*

(a) R is a 1-dimensional, Noetherian, Hilbert domain.

(b) Let $\bar{R}$ be the completion of R in the (x, y)-adic topology. $\bar{R}$ is not a domain, since it is isomorphic to $K[[X, Y]]/(Y^2 - X^3 - X^2)$, and $Y^2 - X^3 - X^2$ factors in $K[[X, Y]]$.

(c) This example shows that the completion of a domain need not be a domain.

(d) This example is from [Nagata 1962, pp. 55–56].

□ □ □ □ □

Example 104

Let $R = K[X, Y]/(f)$, where K is an algebraically closed field with char$(K) = 0$ *and $f = X(X^2 + Y^2) + (X^2 - Y^2)$. Let x, y be the images of X, Y, respectively.*

(a) R is a 1-dimensional, Noetherian, Hilbert domain.

(b) Let $M = (x, y)$. Let $\bar{R}$ be the completion of R in the M-adic topology. Just as in Example 103, $\bar{R}$ is not a domain.

(c) This example is from [Bourbaki 1972, p. 249, Exercise 15].

□ □ □ □ □

Example 105

Let $R = \mathbf{Z}_S$, where S is the complement of $(2) \cup (3)$.

(a) R is a principal ideal domain having precisely 2 maximal ideals, namely (2) and (3).

(b) The Jacobson radical of R is $I = (2) \cap (3) = (6)$. Let $\bar{R}$ be the completion of R in the I-adic topology. $\bar{R}$ is not a domain. In fact, if R_1 and R_2 are the completions of $\mathbf{Z}_{(2)}$ and $\mathbf{Z}_{(3)}$, respectively, then $\bar{R} = R_1 \oplus R_2$.

(c) It is also true that $\bar{R}$ is isomorphic to $R[[X]]/(X - 6)$. This demonstrates the interesting fact that $X - 6$ can be factored in $R[[X]]$, but not in $R[X]$.

In $R[[X]]$, $X - 6 = (-1)(2 - X - X^2 + \ldots)(3 + X + 2X^2 + \ldots)$.

(d) Although $\overline{R}$ is not a domain, it is still Noetherian, Cohen–Macaulay, regular, and 1-dimensional.

(e) This example is from [Nagata 1962, pp. 55–56].

Example 106

Let $T = K[[X, Y]]$, where K is a field with $\mathrm{char}(K) \neq 2$. Let $w \in K[[X]]$ be $w = a_1 X + a_2 X^2 + a_3 X^3 + \ldots$, such that w is transcendental over $K(X)$. Let $u_1 = u = (Y + w)^2$, and for $i > 1$, let $u_{i+1} = X^{-1}(u - (Y + a_1 X + a_2 X^2 + \ldots + a_{i-1} X^{i-1})^2)$. Let $R_0 = K[X, Y, u_1, u_2, u_3, \ldots]$. Let $M = (X, Y, u_1, u_2, u_3, \ldots)$. Let $R_1 = (R_0)_M$. Let $R = R_1[Z]/(Z^2 - u)$.

(a) Clearly, R_1 is a quasi-local domain containing $K[X, Y]$ with M_M lying over (X, Y) in $K[X, Y]$. Therefore, the completion of R_1 is $K[[X,Y]]$ (since it is certainly contained in $K[[X, Y]]$).

(b) Thus R_1 is actually 2-dimensional.

(c) In fact, R_1 is Noetherian, and therefore is regular, since its completion is regular.

(d) Thus R is 2-dimensional and Noetherian. Since u is not a square in R_1, R is an integrally closed domain. R is also local, with the image of $(M, Z^2 - u)$ as its maximal ideal.

(e) The completion of R is $K[[X, Y]][Z]/(Z^2 - u)$, which is not a domain, since $Z^2 - u$ factors in $K[[X, Y]][Z]$, namely as $(Z - Y - w)(Z + Y + w)$. Thus the completion of R has two distinct minimal prime ideals.

(f) Thus R is a 2-dimensional, local, integrally closed domain whose completion is not a domain.

(g) This example is from [Nagata 1962, pp. 209–211].

Example 107

Let $K_0 = \mathbf{Z}/(p)$, where p is a prime integer. Let $K = K_0(X_1, X_2, X_3, \ldots)$. Let $T = K[[Y]]$. Let $R = K^p[[Y]] \otimes_{K^p} K$. Let $c = X_1 Y + X_2 Y^2 + X_3 Y^3 + \ldots$. Let $R_1 = R[c]$.

(a) R is a Noetherian valuation domain whose completion is T.

(b) R_1 is a 1-dimensional Noetherian domain. If $d = c^p$, we can consider R_1 as $R[W]/(W^p - d)$. Thus R_1 is local, with the image of $(Y, W^p - d)$ as its maximal ideal. Notice that $W^p - d$ is irreducible over R.

(c) The completion of R_1 is therefore isomorphic to $T[W]/(W^p - d)$. However, $W^p - d$ factors over $T[W]$, so in the completion of R_1, the image of $W - c$ is nilpotent, i.e., $(W - c)^p = W^p - d = 0$.

(d) Thus R_1 is a 1-dimensional local domain whose completion contains nonzero nilpotent elements.

(e) This example is from [Nagata 1962, p. 206].

Example 108

Let $K_0 = \mathbf{Z}/(2)$. Let $K = K_0(X_1, X_2, X_3, \ldots)$. Let $T_0 = K[[Y, Z]]$. Let $R = K^2[[Y, Z]] \otimes_{K^2} K$. Let $d = (X_1 Y + X_3 Y^2 + X_5 Y^3 + \ldots) + (X_2 Z + X_4 Z^2 + X_6 Z^3 + \ldots)$. Let $R_1 = R[d]$.

(a) R is a 2-dimensional, regular, local domain whose completion is T_0.

(b) R_1 is integral over R, since $d^2 \in R$, so R_1 is 2-dimensional and Noetherian.

(c) If $c = d^2$, then R_1 is isomorphic to $R[W]/(W^2 - c)$. R_1 is local, with the image of $(Y, Z, W^2 - c)$ as its maximal ideal. Let T be the completion of R_1.

(d) T has nonzero nilpotent elements, since it is isomorphic to $T_0[W]/(W^2 - c)$ and $(W - d)^2 = W^2 - c = 0$.

(e) R_1 is actually integrally closed, although this is not easy to prove. Thus R_1 is an integrally closed domain whose completion has nonzero nilpotent elements.

(f) This example is from [Nagata 1962, p. 208].

Example 109

Let $R = \mathbf{R}[X, Y]$. Let $P = (X^2 + Y^2 - 1)$. Let T be the completion of R in the P-adic topology.

(a) R is a 2-dimensional, regular, factorial domain.

(b) R/P is a Dedekind domain, with $\mathrm{Cl}(R/P)$ isomorphic to $\mathbf{Z}/2\mathbf{Z}$. (See Example 55.)

(c) T is a 2-dimensional regular domain with P contained in its Jacobson radical. T is not factorial; in fact, $\mathrm{Cl}(T)$ is isomorphic to $\mathbf{Z}/2\mathbf{Z}$.

(d) This seeming coincidence, that $\mathrm{Cl}(R/P)$ and $\mathrm{Cl}(T)$ are isomorphic, is not a coincidence. Under appropriate conditions on a domain S and a prime ideal Q, $\mathrm{Cl}(\overline{S})$ is isomorphic to $\mathrm{Pic}(S/Q)$, where $\overline{S}$ is the completion of S in the Q-adic topology and $\mathrm{Pic}(S/Q)$ is the Picard group of S/Q. (The *Picard group* of a domain S is the subgroup of $\mathrm{Cl}(S)$ generated by the classes of invertible ideals. For a Dedekind domain S, $\mathrm{Cl}(S) = \mathrm{Pic}(S)$.)

(e) This example shows that the completion of a factorial domain need not be factorial. See also Example 95.

(f) This example is from [Fossum 1973, p. 113].

Example 110

Let $L_1 = K(X)$ and $L_2 = K(Y)$, where K is a field. Let $T = L_1 \otimes_K L_2$.

(a) L_1 and L_2 are both fields and T can be considered as a subring of $K(X, Y)$.

(b) T is not all of $K(X, Y)$. In fact, T is not a field. Specifically, the element $X + Y$, or more precisely $X \otimes 1 + 1 \otimes Y$, is not invertible in T. Elements of $K[X, Y]$ have inverses in T if they are of the form fg, where f and g are nonzero elements of $K[X]$ and $K[Y]$, respectively.

(c) Since T lies between $K(X)[Y]$ and $K(X, Y)$, it can be shown that T is actually a principal ideal domain. This example shows that the tensor product of two fields may be a domain without being a field.

Example 111

Let $K = K_0(X^2)$, where K_0 is a field. Let L_1 and L_2 be two isomorphic copies of $K_0(X)$. Let $T = L_1 \otimes_K L_2$.

(a) T is isomorphic to $K[Y, Z]/(Y^2 - X^2, Z^2 - X^2)$. Let y, z be the images of Y, Z, respectively. T is therefore a 1-dimensional Noetherian ring.

(b) T is not a domain, since $(y - z)(y + z) = 0$. If $\mathrm{char}(K_0) = 2$, we have $(y - z)^2 = 0$, so T may even have nilpotent elements.

(c) This shows that the tensor product of two fields may have zero-divisors, or even nilpotent elements. Notice that L_1 (or L_2) is a finite inseparable algebraic extension of K. It is not essential that L_1 and L_2 are isomorphic; we could easily enlarge each by a finite algebraic extension to obtain nonisomorphic fields L_3 and L_4 such that $L_3 \otimes_K L_4$ is not a domain.

Example 112

Let $K = K_0(X^2)$, where K_0 is a field with $\mathrm{char}(K_0) = 2$. Let $R = K[Y, Z]/(Y^2 + X^2Z + Z^3)$. Let $L = K_0(X)$. Let $T = R \otimes_K L$.

(a) R is a Dedekind domain, i.e., a 1-dimensional, Noetherian, integrally closed domain.

(b) T is isomorphic to $L[Y, Z]/(Y^2 + X^2Z + Z^3)$. Let y, z be the images of Y, Z, respectively. We have $y^2 = X^2z + z^3 = z(X + z)^2$. The element $Y/(X + z)$ is in the quotient field of T and is integral over T, but is not in T. Thus T is not integrally closed.

(c) T is a 1-dimensional Noetherian domain.

(d) R is an example of a regular domain which is not geometrically regular or geometrically normal. Notice that L is a purely inseparable finite algebraic extension of K.

Example 113

Let K_0 be a field with char$(K_0) = p > 2$. *Let $K = K_0(X^p)$. Let $L = K_0(X)$. Let $R = K[Y, Z]/(f)$, where $f = Y^2 - Z^p - X^p$. Let $T = R \otimes_K L$.*

(a) R is a Dedekind domain.

(b) T is isomorphic to $L[Y, Z]/(f)$, so it is 1-dimensional and Noetherian. Let y, z be the images of Y, Z, respectively. We have $y^2 = z^p + X^p = (z + X)^p$. Since $p \neq 2$, the polynomial f does not factor in $L[Y, Z]$, so T is a domain.

(c) T is not integrally closed, since the element $y/(z + X)$ is in the quotient field of T and is integral over T, but is not in T.

(d) R is an example of a regular domain which is not geometrically regular or geometrically normal. Notice the similarities between this example and Example 112. As in that example, L is a purely inseparable finite algebraic extension of K.

(e) This example is from [Bourbaki 1972, p. 360, Exercise 23] and from [Dieudonne 1967, p. 99].

Example 114

Let K_0 be a field with char$(K_0) = 2$. *Let $K = K_0(X^2)$ and $L = K(X)$. Let $R_0 = K[U, V, W]/(f)$, where $f = U^3 + V^3 + W^2 + X^2$. Let u, v, w be the images of U, V, W, respectively. Let $M = (u, v, w^2 + X^2)$, a maximal ideal in R_0. Let $R = (R_0)_M$. Let $T = R \otimes_K L$.*

(a) R is clearly a 2-dimensional local ring. Since f is irreducible in $K[U, V, W]$, R is a domain. Since $f \in M$, but $f \notin M^2$, R is actually a regular domain.

(b) T is a 2-dimensional semi-local domain. This is easy to see, since T is integral over R and the polynomial f is irreducible over $L[U, V, W]$.

(c) T is not regular. It contains a maximal ideal $N = (u, v, w + X)$, which lies over M. Since $f = u^3 + v^3 + (w + X)^2$, f is in the pre-image of N^2 (where we consider $T = (L[U, V, W]/(f))_N$); so T is not regular. It is integrally closed, however.

(d) T is thus a regular local domain which is not geometrically regular. I don't know whether R is geometrically normal. Since a 1-dimensional Noetherian domain is regular if and only if it is integrally closed, one must consider rings of dimension at least 2 to find a regular domain which is geometrically normal without being geometrically regular.

Example 115

Let $S_1 = K[X_1]$, where K is a field. For $i > 1$, let $S_i = S_{i-1}[X_i]/(X_i^2 - X_{i-1})$. For each i, let x_i be the image of X_i, and let $R_i = (S_i)_M$, where $M = (x_i)$.

(a) For each i, R_i is a Noetherian valuation domain. For $i > 1, R_i$ is integral over R_{i-1}.

(b) We have $R_1 \subseteq R_2 \subseteq R_3 \subseteq \ldots$, an ascending chain of domains. Let $T =$ the union of this chain. T is a 1-dimensional non-Noetherian valuation domain, whose maximal ideal is generated by the images of $x_1, x_2, x_3, \ldots$.

(c) Thus the union of an ascending chain of Noetherian factorial domains does not have to be Noetherian or factorial. Notice that the irreducible elements of R_i (there actually only is one) do not remain irreducible in R_{i+1}.

Example 116

For $i > 0$, let $R_i = K[X_1, X_2, \ldots, X_i]$, where K is a field.

(a) For $i > 0$, R_i is an i-dimensional, regular, factorial domain.

(b) We have $R_1 \subseteq R_2 \subseteq R_3 \subseteq \ldots$, an ascending chain of domains. Let $T =$ the union of this chain. $T = K[X_1, X_2, X_3, \ldots]$, so T is an infinite-dimensional factorial domain. (In fact, it is the ring of Example 1.)

(c) Notice that the irreducible elements of R_i remain irreducible in R_{i+1}. For $i > 0$, R_{i+1} and R_i together satisfy LO and GD, so T satisfies LO and GD with respect to R_i, for each i. Specifically, if $P = P_1$ is a prime ideal of R_1, we can find, for each $i > 1$, a prime ideal P_i of R_i, lying over P_{i-1}. Then $P_1 \subseteq P_2 \subseteq P_3 \subseteq \ldots$, and if $Q =$ the union of this chain, then Q is a prime ideal of T lying over P.

Example 117

For $i > 0$, let $K_i = K(X_1, X_2, \ldots, X_i)$, where K is a field. Let $R_1 = K[X_1]$, and for $i > 1$, let $R_i = K_{i-1}[X_i]$.

(a) For $i > 0$, R_i is a principal ideal domain.

(b) We have $R_1 \subseteq R_2 \subseteq R_3 \subseteq \ldots$. Let L = the union of this chain. Then $L = K(X_1, X_2, X_3, \ldots)$, a field. Thus the union of an ascending chain of 1-dimensional domains may be a field, i.e., 0-dimensional. Since R_{i+1} contains the quotient field of R_i, all this example really involves is the fact that L is the union of an ascending chain of subfields $K \subseteq K_1 \subseteq K_2 \subseteq \ldots$.

(c) By adjoining more indeterminates, we could construct an m-dimensional domain which is the union of an ascending chain of $(m + 1)$-dimensional domains. By grouping the indeterminates differently, such as $K_i' = K(X_1, \ldots, X_{ni})$ and $R_i' = K_i'[X_{ni+1}, X_{ni+2}, \ldots, X_{ni+n}]$, we can represent the field L as the union of an ascending chain of n-dimensional domains R_i'.

Example 118

For $i > 0$, let $K_i = K(X_1, X_2, \ldots, X_i)$, where K is a field. Let $R_1 = K[X_1, X_2, X_3, \ldots]$. For $i > 0$, let $R_{i+1} = K_i[X_{i+1}, X_{i+2}, X_{i+3}, \ldots]$.

(a) For each $i > 0$, R_i is an infinite-dimensional factorial domain.

(b) We have $R_1 \subseteq R_2 \subseteq R_3 \subseteq \ldots$, an ascending chain of domains. Let L = the union of this chain. Then $L = K(X_1, X_2, X_3, \ldots)$, a field. Thus it is possible for the union of an ascending chain of infinite-dimensional domains to be a field, i.e., 0-dimensional.

(c) Notice that none of the rings R_i are Noetherian, but their union, L, is Noetherian. For each $i > 1$, R_{i+1} is a localization of R_i. Thus this example amounts to a sequence of localizations of R_1, each of which leaves the dimension unchanged, but all of which taken together yield the quotient field of R_1, namely L.

Example 119

For $i > 0$, let $K_i = K(X_1, \ldots, X_i)$, where K is a field. Let $R_1 = K[X_1, Y_1, \ldots, Y_n]$. For $i > 1$, let $R_i = K_{i-1}[X_i, X_{i+1}, \ldots, X_{2i-1}, Y_1, Y_2, \ldots, Y_n]$.

(a) For $i > 0$, R_i is an $(n + i)$-dimensional, regular, factorial domain.

(b) We have $R_1 \subseteq R_2 \subseteq R_3 \subseteq \ldots$, an ascending chain of domains. Let $T =$ the union of this chain. Then $T = L[Y_1, Y_2, \ldots, Y_n]$, where L is the field $L = K(X_1, X_2, X_3, \ldots)$. T is an n-dimensional, regular, factorial domain.

(c) Here the set $\{\dim(R_i)\}$ is unbounded, yet the union of the rings R_i is of a dimension smaller than $\dim(R_i)$ for any i.

Example 120

For $m > 0$, let $S_m = \mathbf{Z}[X/m!]$, and let $R_m = (S_m)_N$, where in each ring S_m, N is the maximal ideal $N = (2, X/m!)$.

(a) For $m > 0$, R_m is a 2-dimensional, regular, local domain.

(b) We have $R_1 \subseteq R_2 \subseteq R_3 \subseteq \ldots$, an ascending chain of domains. Let $T =$ the union of this chain. Then T is a 2-dimensional valuation domain of the form $T = \mathbf{Z}_{(2)} + X(\mathbf{Q}[X])_{(X)}$.

(c) Thus the union of an ascending chain of regular local domains does not have to be factorial or Noetherian or even completely integrally closed, although it must be quasi-local and integrally closed.

Example 121

Let R be any ring. For $i > 0$, let $R_i = R[X_i, X_{i+1}, X_{i+2}, \ldots]$.

(a) Each of the rings R_i is infinite-dimensional and non-Noetherian, regardless of the structure of R.

(b) We have $R_1 \supseteq R_2 \supseteq R_3 \supseteq \ldots$, a descending chain of rings, and $R =$ the intersection of this chain. Thus any ring whatsoever is the intersection of a descending chain of infinite-dimensional non-Noetherian rings.

Example 122

Let R be any ring. For $i > 0$, let $R_i = R[X^{2^{i-1}}]$.

(a) For $i > 0$, R_i is integral over R_{i+1}. The rings R_i are all isomorphic; if $\dim(R) = n$, then $\dim(R_i) \leqslant 2n + 1$.

(b) We have $R_1 \supseteq R_2 \supseteq R_3 \supseteq \ldots$, a descending chain of rings, and $R =$ the intersection of this chain. Thus any ring whatsoever is the intersection of a descending chain of integral extensions.

(c) Nonetheless, none of the rings R_i are integral over R.

Example 123

Let p_1, p_2, p_3, ... be the sequence of positive prime integers. Let $T = \mathbf{Z}[X_1, X_2, X_3, \ldots]/(X_i^2 - p_i$ for all i). Let s_i be the image of X_i, for all i. For $m > 0$, let $R_m = \mathbf{Z}[s_m, s_{m+1}, s_{m+2}, \ldots]$.

(a) For $m > 0$, R_m is a 1-dimensional domain. I don't know if it is Noetherian or integrally closed, for any m; the latter at any rate seems unlikely.

(b) For $m > 0$, R_m is an integral extension of R_{m+1}, so we have $R_1 \supseteq R_2 \supseteq R_3 \supseteq \ldots$, a descending chain of domains.

(c) We have $\mathbf{Z} = \cap R_m$. Unlike the situation of Example 122, each of the rings R_m is integral over $\mathbf{Z}$.

Example 124

Let K be a field. Let $K_1 = L = K(X)$ and for $i \geqslant 1$, let $K_i = K(X^{2^{i-1}})$. Let $R_1 = K_1[[Y]]$ and for $i > 1$, let $R_i = K_i + Y \cdot L[[Y]]$.

(a) R_1 is a Noetherian valuation domain. For $i > 1$, R_i is a 1-dimensional, local domain. To see this, notice that L is a finite-dimensional algebraic extension of K_i. Therefore R_1 is a finitely generated integral extension of R_i.

(b) We have $R_1 \supseteq R_2 \supseteq R_3 \supseteq \ldots$, a descending chain of domains. Let $T =$ the intersection of this chain. Then $T = K + Y \cdot L[[Y]]$. T is a 1-dimensional, integrally closed, quasi-local, non-Noetherian domain. None of the rings R_i are integral over T, since L is a purely transcendental extension of K.

(c) Thus the intersection of a descending chain of Noetherian rings does not have to be Noetherian.

Example 125

Let K be a field and let n be a positive integer. Let $T = K[X_1, \ldots, X_n]$. Let G be the group of all permutations on n elements; G is called the symmetric *group. Each permutation $g \in G$ acts on $X_1, \ldots, X_n$ by $g(X_i) = X_k$, where $k = g(i)$, and this induces an automorphism on T, also denoted by g. G can therefore be considered as a group of automorphisms on T. Let $R = T_G$.*

(a) T is an n-dimensional, regular, factorial domain.

(b) $R = K[s_1, s_2, \ldots, s_n]$, where $s_1, \ldots, s_n$ are the n elementary symmetric functions of $X_1, \ldots, X_n$. (For instance, $s_1 = X_1 + \ldots + X_n$ and $s_n = X_1 X_2 \ldots X_n$.)

(c) Since $s_1, \ldots, s_n$ are algebraically independent over K, R is just a polynomial ring in n variables over K; in fact, R is isomorphic to T. Since each element of G is of finite order, T is integral over R.

(d) This example is from [Samuel 1969]. (The elementary symmetric functions are of course well known.)

Example 126

Let K be a field of characteristic $\neq 2$. Let n be a positive integer. Let $T = K[X_1, \ldots, X_n]$. Let G be the group of all even permutations on n elements; G is called the alternating group. *Each permutation* $g \in G$ *acts on* $X_1, \ldots, X_n$ *by* $g(X_i) = X_k$, *where* $k = g(i)$, *and this induces an automorphism on T, also denoted by g. G can therefore be considered as a group of automorphisms on T. Let* $R = T_G$.

(a) T is an n-dimensional, regular, factorial domain.

(b) $R = K[s_1, \ldots, s_n, D]$, where $s_1, \ldots, s_n$ are the n elementary symmetric functions of $X_1, \ldots, X_n$ and $D =$ the product of all $(X_i - X_j)$ with $i < j$. Since R is an affine ring, it is certainly Noetherian. Since T is integrally closed and G is a finite group, R is integrally closed.

(c) Since D^2 is left fixed by all permutations whether even or odd, R is a finite integral extension of the ring of Example 125.

(d) I do not know whether R is factorial.

(e) This example is from [Nagata 1977, pp. 45–53].

Example 127

Let $L = K(X_1, X_2, \ldots, Y_1, Y_2, \ldots)$, *where K is the field of two elements. Let* $T = L[[W, Z]]$.

(a) T is a 2-dimensional, regular, local domain.

(b) Let s be the unique automorphism of T such that $s(W) = W$, $s(Z) = Z$, $s(X_i) = X_i + Z(X_{i+1}W + Y_{i+1}Z)$, and $s(Y_i) = Y_i + W(X_{i+1}W + Y_{i+1}Z)$, for each $i > 0$.

(c) Let R be the subring of T consisting of all elements left fixed by s. Since s^2 is the identity automorphism, T is integral over R. Specifically, each $t \in T$ is a root of a monic quadratic polynomial over R.

(d) Thus R is 2-dimensional and quasi-local.

(e) Let L_0 be the quotient field of R. Then $R = T \cap L_0$, so R is a Krull domain. I don't know whether R is factorial or regular or complete. (T is complete.)

(f) For $i > 0$, let $b_i = WX_i + ZY_i$, then $s(b_i) = b_i$, so $b_i \in R$. The ideal $(b_1, b_2, b_3, \ldots)$ is not finitely generated over R, so R is not Noetherian. Thus T is a Noetherian Krull domain which is integral over R, which is Krull but not Noetherian.

(g) Obviously, R contains $K[[W, Z]]$, but R also contains the elements

$c_i = X_i s(X_i) = X_i^2 + Z X_i b_{i+1}$ and $d_i = Y_i s(Y_i) = Y_i^2 + W Y_i b_{i+1}$, which are units in T and in R.

(h) This example is from [Nagarajan 1968]. Notice the resemblance to Example 57.

Example 128

Let $R_0 = \mathbf{Z}[X_1, X_2, \ldots, Y_1, Y_2, \ldots]$. *Let* $K = (R_0)_{(2)}$. *Let* $T = K[[U, V]]$.

(a) K is a Noetherian valuation domain (not a field).

(b) T is a 3-dimensional, regular, factorial domain.

(c) Let s be the unique automorphism of T such that $s(U) = -U$, $s(V) = V$, $s(X_i) = -X_i + b_{i+1} U$, and $s(Y_i) = Y_i + b_{i+1} V$, for each $i > 0$, where $b_i = X_i V + Y_i U$, for such i. Notice that for such i, $s(b_i) = -b_i$.

(d) Let R be the subring of T consisting of all elements left fixed by s. Since s^2 is the identity automorphism, T is integral over R.

(e) Thus R is 3-dimensional. Since T is integrally closed and s is of finite order, R is integrally closed.

(f) R is not Noetherian; this is easy to see, since $R/(2)$ and $T/(2)$ are the rings of Example 127.

(g) This example is from [Chuang and Lee 1977]. It was invented as a characteristic 0 version of Nagarajan's example (Example 127).

Chuang and Lee proved that if T is a Noetherian ring containing ½ and s is an automorphism of T of order 2, then the fixed subring R is Noetherian. Notice that in this example, and in Example 127, neither ring T contains ½.

Example 129

Let K *be a field with* char$(K) = p > 0$. *Let* n *be a positive integer. Let* $T = K[X_1, \ldots, X_n]$. *Let* g *be the automorphism of* T *defined by* $g(a) = a$ *for* $a \in K$, $g(X_n) = X_n$, *and* $g(X_i) = X_i + X_{i+1}$ *for* $i < n$. *Let* $R = T_g = $ *the subring of* T *consisting of all elements left fixed by* g.

(a) T is an n-dimensional, regular, factorial domain.

(b) Since g is of finite order, T is integral over R. Therefore $\dim(R) = n$. Since T is finitely generated as an algebra over K, it is finitely generated as an algebra over R; so T is finitely generated as a module over R. Therefore, R is Noetherian since T is Noetherian, by a theorem of Eakin. Since T is integrally closed and g is of finite order, R is integrally closed, for any n.

(c) Suppose $n = 2$. Let $u = (X_1 + X_2)(X_1 + 2X_2)\ldots(X_1 + pX_2)$. Then $g(u) = u$ and $R = K[u, X_2]$. Since u and X_2 are algebraically independent over K, R is factorial and regular.

(d) Suppose $p = 2$ and $n = 3$. Let $u_1 = X_3$, $u_2 = X_2 g(X_2)$, $u_3 = X_1 g(X_1) \cdot g^2(X_1)g^3(X_1)$, and $u_4 = X_1 X_3(X_1 + X_3) + X_2^2(X_2 + X_3)$. Then $g(u_i) = u_i$ for each i, and $R = K[u_1, u_2, u_3, u_4]$. In this case, R is factorial and Cohen–Macaulay.

(e) Suppose $p = 2$ and $n = 4$. As in the earlier cases, but at much greater effort, it is possible to define elements u_1, u_2, u_3, u_4, x, and y, such that R is the integral closure of $K[u_1, u_2, u_3, u_4, x, y]$ in its quotient field. Since $\dim(R) = 4$, there are, of course, various algebraic identities satisfied by u_1, u_2, u_3, u_4, x, and y. In this case, R is factorial, but not Cohen–Macaulay.

(f) If the ring of part (e) is localized appropriately, we have a 4-dimensional, local, factorial domain which is not Cohen–Macaulay. It has been shown that the completion of this local ring is still factorial.

(g) This example is from [Bertin 1967]. See also [Fossum 1973, pp. 86–88], [Fossum and Griffith 1975], and item #18 in the Further References section.

□ □ □ □ □

Example 130

Let $R = \mathbf{Z}[X]/(2X, X^2)$.

(a) $\mathbf{Z}$ is a principal ideal domain.

(b) R is an integral extension of $\mathbf{Z}$, so R is 1-dimensional. Clearly, it is also Noetherian. The unique minimal prime ideal of R is nilpotent, generated by the image of X.

(c) The prime ideals of R and of $\mathbf{Z}$ are in one-to-one correspondence. Each ring has a unique minimal prime ideal and countably many maximal ideals, each of height 1. Thus GD holds for the pair $\mathbf{Z}, R$.

(d) Since there are zero-divisors of $\mathbf{Z}$ on R, R is definitely not a flat $\mathbf{Z}$-module.

(e) If we have two rings S and T, with S a subring of T and T a flat S-module, GD must hold for the pair S, T. (See [Matsumura 1970, p. 33] for a proof.) This example demonstrates that the converse is false.

(f) An S-module M is flat if and only if it is flat locally, i.e., if and only if for each prime ideal P of S, the S_P-module $M_P = M \otimes S_P$ is flat over S_P. In this example, R passes this test for every prime ideal of $\mathbf{Z}$ except (2).

□ □ □ □ □

Example 131

Let S and T be any two rings. Let $R = S \oplus T$.

(a) S can be identified with $S \oplus (0)$, an ideal of R. S is thus an R-module, and since S is a summand of R, a free R-module, S is a projective R-module.

(b) Since S is projective, it is flat. However, S is not faithfully flat, since $S \otimes_R (R/(S \oplus (0)) = 0$.

(c) If $M = R \oplus R \oplus R \oplus \ldots$, then $S \oplus M$ is isomorphic to M, since $S \oplus M = S \oplus ((S \oplus T) \oplus (S \oplus T) \oplus ..)$, which is isomorphic to $S \oplus (T \oplus S) \oplus (T \oplus S) \oplus \ldots$, which is isomorphic to $(S \oplus T) \oplus (S \oplus T) \oplus \ldots = M$. Thus S is a direct summand of a free module, with a free complement.

(d) By the same device, any projective module can be shown to have a free complement, as a summand of a free module. To require the complement to be finitely generated and free is a much stronger condition.

(e) This example is from [Matsumura 1970, p. 17].

□ □ □ □ □

Example 132

Let $R = \mathbf{Z}[\sqrt{-5}]$.

(a) R is a Dedekind domain.

(b) Let $P = (3, 2 + \sqrt{-5})$. P is a prime ideal of height 1, and is invertible, since every nonzero ideal of R is invertible.

(c) Since P is invertible, it is projective as an R-module. However, P is not principal, so it is not free. (In a domain, an ideal is free if and only if it is principal and invertible if and only if it is projective.) Thus P is a projective R-module which is not free.

Example 133

Let S be the set of all continuous functions from the closed interval $[0,1]$ into the real numbers. S is a ring under pointwise addition and multiplication. Let J be the set of all $f \in S$ such that for some $e > 0$, f is identically 0 on the interval $[0,e]$.

(a) J is an ideal of S.

(b) J is a faithful S-module, i.e., the annihilator of J is just (0).

(c) J is projective. To prove this, we shall see that J is a direct summand of $M = S \oplus S \oplus S \oplus \ldots =$ the free S-module on countably infinitely many generators. Elements of M are sequences $(a_0, a_1, a_2, \ldots)$, with all but finitely many $a_i = 0$.

For $n > 0$, let $f_n(t) = 0$ for $0 \leqslant t \leqslant 1/(n + 1)$, $f_n(t) = n(n + 1)(t - 1/(n + 1))$ for $1/(n + 1) \leqslant t \leqslant 1/n$, and $f_n(t) = 1$ otherwise. Let $g_0 = \sqrt{f_1}$ and for $n > 0$, let $g_n = \sqrt{f_{n+1} - f_n}$. Then for $n \geqslant 0$, $g_0^2 + g_1^2 + \ldots + g_n^2 = f_{n+1}$, and each $g_i \in J$.

Let $p : M \to J$ be $p(a_0, a_1, \ldots) = a_0 g_0 + a_1 g_1 + \ldots$, and let $q : J \to M$ be $q(h) = (h g_0, h g_1, \ldots)$. Then for $h \in J$, we have $p(q(h)) = h$. This gives a split short exact sequence $0 \to J \xrightarrow{q} M \to M/q(J) \to 0$, so J is a summand of M, and J is projective.

(d) J is an example of a projective ideal which is not finitely generated. In a domain, any projective ideal must be finitely generated.

(e) This example is from [Cohen 1969]; also see Example 180.

Example 134

Let K be a field. Let $R_0 = K[X^2, X^3]$ and let $M_0 = (X^2, X^3)$. M_0 is a maximal ideal of R_0. Let $R = (R_0)_{M_0}$.

(a) R is a 1-dimensional local domain and is therefore a G-domain.

(b) There is a homomorphism from $R[Y]$ onto $R[X^{-1}] = K(X)$ sending Y to X^{-1}. Let Q be the kernel of this mapping. Then Q is an ideal of $R[Y]$, and Q is not principal, since $X^2 Y^2 - 1$ and $X^3 Y^3 - 1$ are in Q and their g.c.d. is 1.

(c) Q is actually invertible (and therefore projective). To show that Q is invertible, we observe that it is finitely generated (since R and $R[Y]$ are Noetherian) and that Q_N is principal in $R[Y]_N$ for each maximal ideal N of $R[Y]$. Thus Q is a projective $R[Y]$-module which is not free.

(d) This example is from [Kaplansky 1974b, p. 75, Exercises 20–21].

□ □ □ □ □

Example 135

Let Z be the ring of integers and let Q be the field of rational numbers.

(a) Q is, of course, the quotient field of $\mathbf{Z}$. Since $\mathbf{Z}$ is a domain and $\mathbf{Q}$ is a torsion-free divisible $\mathbf{Z}$-module, $\mathbf{Q}$ is an injective $\mathbf{Z}$-module. In fact, $\mathbf{Q}$ is the injective envelope of $\mathbf{Z}$ as a $\mathbf{Z}$-module.

(b) Consider $\mathbf{Q}/\mathbf{Z}$ as a $\mathbf{Z}$-module. $\mathbf{Q}/\mathbf{Z}$ is divisible, but every element is of finite order. Since $\mathbf{Z}$ is a Dedekind domain, divisible $\mathbf{Z}$-modules are injective, so $\mathbf{Q}/\mathbf{Z}$ is injective.

(c) For each prime integer p, let H_p be the subgroup of $\mathbf{Q}/\mathbf{Z}$ consisting of all elements annihilated by some power of p. Then H_p is the injective envelope of $\mathbf{Z}/(p)$, and $\mathbf{Q}/\mathbf{Z}$ is isomorphic to the direct sum of the subgroups H_p for all primes p.

□ □ □ □ □

Example 136

Let R be a domain with quotient field K.

(a) K is a torsion-free divisible R-module, so K is injective as an R-module.

(b) If R is Dedekind, then K/R is an injective R-module. If R is not Dedekind, then K/R is still divisible as an R-module, but may fail to be injective.

□ □ □ □ □

Example 137

Let K be any field. Let R be the group algebra of $\mathbf{Z}/(2)$ *over K.*

(a) R can be described as $K[a]$, where $a^2 = 1$. Thus R is a 0-dimensional Noetherian ring with at most two prime ideals, $(a + 1)$ and $(a - 1)$. (Of course, if char$(K) = 2$, these are identical.)

(b) Since R is isomorphic to $K[X]/(X^2 - 1)$, where $K[X]$ is a regular domain and $X^2 - 1$ is a non-zero-divisor, R is Gorenstein. Therefore id$(R) = \dim(R) = 0$, so R is injective as a module over itself, i.e., R is quasi-Frobenius.

(c) In fact, if H is any finite abelian group and T is the group algebra of H over K, then T is quasi-Frobenius. One way to prove this is to observe that any finite abelian group is the direct sum of finitely many cyclic groups. Suppose $H = H_1 \oplus H_2 \oplus \ldots \oplus H_n$, where each group H_i is cyclic. Then T is isomorphic to $K[X_1, \ldots, X_n]/I$, where I is a suitable ideal generated by an R-sequence of length n, and the image of X_i is a generator for H_i. It follows then that R is Gorenstein and $\dim(R) = 0$.

(d) See [Rotman 1970, pp. 91–92] for a proof that the group algebra of any (not necessarily abelian) finite group over any field is quasi-Frobenius.

Example 138

Let K be a field. Let $R = K[X, Y]$ *and let B be the R-module* $B = R/(X)$.

(a) R is a 2-dimensional, regular, factorial domain.

(b) B is not a flat R-module, since a flat module over a domain must be torsion-free, and B, by its construction, has torsion: $XB = 0$.

(c) One of the useful properties of flat modules is that if M is a flat S-module (for some ring S) and I, J are ideals of S, then $(I \cap J)M = IM \cap JM$. This property can also be used to show that B is not a flat R-module.

Let $I = (X + Y)$ and $J = (Y)$ be ideals of R. Then $(I \cap J)B = (XY + Y^2)B = Y^2B$, and $IB \cap JB = (X + Y)B \cap (Y)B = (Y)B$. The sequence $(0) \to I \cap J \to R \to (R/I \oplus R/J) \to (0)$ is exact, but the sequence $(0) \to B \otimes (I \cap J) \to B \otimes R \to B \otimes (R/I \oplus R/J) \to (0)$ is not.

(d) This example is from [Matsumura 1970, p. 23].

Example 139

Let K be a field. Let R = K[X, Y], and let B be the R-module K[X, Y/X].

(a) R is a 2-dimensional, regular, factorial domain.

(b) B is not a localization of R, although B lies between R and its quotient field. (If T is a principal ideal domain, or any QR domain, then any ring lying between T and its quotient field is a localization of T.)

(c) B is obviously a torsion-free R-module, but is not a flat R-module. To see this, let $I = (X)$ and $J = (Y)$, and observe that $B(I \cap J) \neq BI \cap BJ$. By part (c) of Example 138, B is not flat.

(d) This example is from [Matsumura 1970, p. 24].

Example 140

*Let S = **Z**/(2) ⊕ **Z**/(2) ⊕ **Z**/(2) ⊕ . . . be the direct sum of countably infinitely many copies of **Z**/(2). Then S is a ring without* 1. *Let R = S ⊕ **Z**, with the operations (a, m) + (b, n) = (a + b, m + n) and (a, m)(b, n) = (ab + na + mb, mn), for all a, b ∈ S and m, n ∈ **Z**.*

(a) R is a 1-dimensional ring (with 1) which is not Noetherian and not a domain.

(b) Let $f = (0, 2)$. The ideal (f) is flat as an R-module. To see this, one shows that $(f)_M$ is flat as an R_M-module for every maximal ideal M of R, as in Example 130, part (f).

(c) Clearly (f) is not faithfully flat, since $(f) \otimes S = 0$, where S is considered as an R-module, and $S \neq 0$.

(d) In fact, (f) is not projective, either. Thus we have an example of a principal ideal of a ring (with zero-divisors) which is not even projective, much less free. A principal ideal of a domain is clearly a free module over the domain.

(e) This example is from [Vasconcelos 1969].

Example 141

Let R = K[X, Y], where K is a field. Let M = (X, Y).

(a) R is a 2-dimensional, regular, factorial domain.

(b) M is an ideal of the domain R, so M is torsion-free as an R-module.

(c) M is not flat. To see this, consider the exact sequence $(0) \to M \to R \to R/M \to (0)$. The sequence $(0) \to M \otimes M \to R \otimes M \to (R/M) \otimes M \to (0)$ is not exact, since $M \otimes M$ is isomorphic to M^2, $R \otimes M$ is isomorphic to M, and $(R/M) \otimes M = (0)$. M is not isomorphic to M^2, so the sequence is not exact.

(d) Notice that since R is not Prüfer, there must be torsion-free R-modules which are not flat. M is one such module.

(e) This example is from [Bourbaki 1972, p. 41, Exercise 3].

Example 142

*Let A be the **Z**-module A = **Z**/(2). Let S = **Z**/(4).*

(a) A can be considered as either a $\mathbf{Z}$-module or as an S-module. $\mathbf{Z}$ is a principal ideal domain. S is a 0-dimensional local ring.

(b) Since $\mathrm{Gd}(\mathbf{Z}) = 1$, $\mathrm{pd}_{\mathbf{Z}}(A)$ must be either 0 or 1. Since a projective module over a principal ideal domain must be free, and A is certainly not a free $\mathbf{Z}$-module, A is not projective, so $\mathrm{pd}_{\mathbf{Z}}(A)$ is not 0, and must be 1. A projective resolution for A might be $(0) \to (2) \to \mathbf{Z} \to A$.

(c) Since $\dim(S) = 0$ and S is not a field, S is not regular, so $\mathrm{Gd}(S) = \infty$. Since S is local and projective modules over a local ring are free, and A is certainly not a free S-module, A is not projective. Therefore $\mathrm{pd}_S(A)$ is not 0. In fact, $\mathrm{pd}_S(A) = \infty$. One projective resolution for A is $\ldots \to S \to S \to S \to A$, where each map $S \to S$ is just multiplication by 2.

(d) Thus the same abelian group, A, is of projective dimension 1 over $\mathbf{Z}$ and ∞ over S. The projective dimension of an abelian group as a module over a given ring depends on the ring as well as on the abelian group.

(e) This example is from [Cohen 1969] and from [Kaplansky 1974a, p. 169].

Example 143

Let K be a field. Let $S_0 = K[[$all $X^{p/q}$ for rational numbers $p/q \geqslant 0]]$. Let S be the subring consisting of all elements $a_0 + a_1 X^{n_1} + a_2 X^{n_2} + \ldots$ for which the sequence n_1, n_2, n_3, $\ldots$ is well-ordered and increasing. Let $R = S/($all $X^{p/q}$ with $p/q > 1)$.

(a) S is a 1-dimensional non-Noetherian valuation domain.

(b) R is a 0-dimensional quasi-local ring whose ideals are linearly ordered.

(c) It can be shown that $\mathrm{Gd}(R) = \infty$.

(d) Consider the principal ideal $I = (x)$, where x is the image of X in R. I is not a free R-module; in fact, $\mathrm{pd}(I) = 2$. To show this, we construct a (minimal) projective resolution for I. Let $R \to I$ be the map which is multiplication by x. The kernel of this map is the ideal $J = (x, x^{1/2}, x^{1/3}, x^{1/4}, \ldots)$. Let M be a free R-module on countably infinitely many generators Y_1, Y_2, Y_3, $\ldots$. Now there is a unique map of M onto J for which the image of Y_i is $x^{1/i}$, for each $i > 0$. If N is the kernel of this map, we have an exact sequence $(0) \to N \to M \to R \to I \to (0)$. It can be shown that N is a free R-module, generated by the elements $w_n = Y_n - x^{e_n} Y_{n+1}$, where $e_n = 1/n - 1/(n + 1)$, for $n > 0$. Thus $\mathrm{pd}(I) = 2$.

(e) One reason why this example is of interest is that a quasi-local ring of finite global dimension whose ideals are linearly ordered must be a domain. Compare this example to Example 142. It would be interesting to know $\mathrm{pd}(A)$ for some other principal ideal A, say $A = (x^{1/2})$.

(f) This example is from [Osofsky 1968].

□ □ □ □ □

Example 144

Let R be an n-dimensional, regular, local ring with maximal ideal M.

(a) R is of course a free R-module. It can be shown, by induction on n, that $\mathrm{pd}(M) = n - 1$. In the case of $n = 1$, this is true since M is principal and therefore free. For the inductive step, one uses various theorems relating $\mathrm{pd}_T(N)$ and $\mathrm{pd}_{T/(t)}(N/tN)$, where N is a T-module and t is not a zero-divisor on either T or N.

(b) By considering the exact sequence $(0) \to M \to R \to R/M \to (0)$, one can show that $\mathrm{pd}(R/M) = n$. Conversely, one can show that a local ring is regular if the projective dimension of its residue field is finite.

(c) For the change-of-rings theorems referred to in (a), see [Kaplansky 1974a]. For the statements in (b), see [Kaplansky 1974a] and [Kaplansky 1974b].

□ □ □ □ □

Example 145

Let $T = K[X]/(X^6)$, where K is a field. Let x be the image of X in T and let $R = K[x^2, x^3]$.

(a) T and R are both 0-dimensional local rings, with T integral over R.

(b) T is Gorenstein, since X^6 is a non-zero-divisor in the Gorenstein domain $K[X]$.

(c) R is not Gorenstein. That is to say, $\text{id}(R)$ is not finite. If it were finite, it would have to be 0; so to show that R is not Gorenstein, it is enough to show that it is not injective as a module over itself. We shall use the definition of injectivity. R is a submodule of the R-module T, and we can map R onto itself by the identity map. There is no map of T into R for which the composite $R \to T \to R$ is the identity map. (Specifically, there is no suitable image for x in R.)

(d) Another way to show that R is not injective over itself is to show that it is not a summand of the R-module T. A module is injective if and only if it is a direct summand of every module containing it.

□ □ □ □ □

Example 146

Let $T = K[X]$ and let $R = K[X^3, X^4, X^5]$, where K is a field.

(a) T is a principal ideal domain and R is a 1-dimensional Noetherian domain.

(b) T is integral over R.

(c) T is regular, and R is Cohen–Macaulay, but R is not Gorenstein. If R were Gorenstein, then $R_0 = R/(X^6)$ would be also since X^6 is a non-zero-divisor in R. To show that R_0 is not Gorenstein, we can use the definition of injectivity, as follows:

Let x be the image of X in R_0. Since R_0 is 0-dimensional, it is enough to show that it is not injective as a module over itself. The ideal (x^8) is a submodule of R_0. Define a map $f: (x^8) \rightarrow R_0$ by $f(x^8) = x^7$. Then there is no way to extend f to a map $R_0 \rightarrow R_0$; specifically, there is no suitable choice for $f(x^3)$.

(d) Thus T is a regular domain, integral over R, and R is not Gorenstein, but R is Cohen–Macaulay. Also see Example 129.

Example 147

Let $R = K[X^3, X^2Y, XY^2, Y^3]$, where K is a field.

(a) R is a 2-dimensional, integrally closed, Noetherian domain. Therefore it is Cohen–Macaulay.

(b) R is not Gorenstein. To show this, we observe that X^6, Y^6 is an R-sequence. If R were Gorenstein, then $T = R/(X^6, Y^6)$ must also be. T is 0-dimensional and local. T is injective if and only if it is injective as a module over itself. To show that it is not injective over itself, we could use the same methods as in Examples 145 and 146.

Example 148

Let $R = \mathbf{Z} \oplus \mathbf{Z} \oplus \mathbf{Z} \oplus \ldots$ be the direct sum of countably infinitely many copies of $\mathbf{Z}$.

(a) R is a ring without 1. That is, there is no element e with $xe = x$ for all $x \in R$.

(b) R does have the property that for each element $y \in R$, there is an element x with $xy = y$. To see this, consider y as a sequence $(b_1, b_2, b_3, \ldots)$, where $b_i = 0$ for all but finitely many i. One choice for x is $x = (a_1, a_2, a_3, \ldots)$, where $a_i = 1$ for all i for which $b_i \neq 0$, and $a_i = 0$ otherwise.

(c) R is 1-dimensional. The minimal prime ideals of R are of the form $P = \mathbf{Z} \oplus \mathbf{Z} \oplus \ldots \oplus \mathbf{Z} \oplus (0) \oplus \mathbf{Z} \oplus \ldots$. Then R/P is isomorphic to $\mathbf{Z}$, which is 1-dimensional.

(d) R is obviously not Noetherian. However, for each prime ideal P of R, R_P is a field or a regular local ring. Thus R is locally a Noetherian factorial domain, but has none of these properties before localization.

Example 149

Let $K = \mathbf{Z}/(p)$, where p is a prime integer. Let $R = K[X, X^{1/2}, X^{1/3}, X^{1/4}, \ldots]$. Let N be the maximal ideal generated by $X, X^{1/2}, X^{1/3}, X^{1/4}, \ldots$. Let $T = R_N$. Let M be the maximal ideal of T.

(a) Consider M as a ring without 1. Then $M^2 = M$.

(b) The only prime ideal of M is (0), unless we consider M itself as a prime ideal. Thus $\dim(M) = 0$. A 0-dimensional "domain without 1" (i.e., a commutative ring with no zero-divisors) is not necessarily a field.

(c) M is not Noetherian, since ideals of T are also ideals of M, and T is not Noetherian.

(d) Let S be the subring of T generated by 1 and M. If K is the field of p elements, as above, then $S = T$ and $\dim(S) = 1 = \dim(M) + 1$. If $K = \mathbf{Q}$ (and the construction is otherwise the same), then $S = \mathbf{Z} + M$ and $\dim(S) = 2 = \dim(M) + 2$.

(e) In any case, S is integral over M. If $K = \mathbf{Q}$, this shows that INC fails for integral extensions (although this does involve assuming that M is itself a prime ideal.)

(f) Although T is a valuation domain, M may not be. For instance, if $K = \mathbf{Q}$, then of the two elements X and $2X$, neither divides the other in M.

(g) Let L be the quotient field of T. Then for any nonzero element t of M, $M[t^{-1}] = L$. Therefore, any nontrivial localization of M is all of L, and the intersection of all nontrivial localizations of M is L, which is much larger than S.

(h) This example is from [Gilmer1972a, p. 270, Exercise 7].

Example 150

Let T = Z[[X]] and let R = (X).

(a) R is a prime ideal of T. We shall consider R as a ring without 1.

(b) Notice that $R^2 \neq R$, and in general, for $x \in R$, there is no $y \in R$ with $xy = x$.

(c) Not every ideal of R is principal. For instance, the ideal $I = (2X, X^3)$ is not principal.

(d) It can be proven (using, say, [Kaplansky 1974b, Theorem 70]) that every proper prime ideal of R is principal; indeed (0) and (pX) are the only proper prime ideals, where p ranges over the prime integers. Thus every prime ideal of R is principal, but not every ideal. This situation is not possible in a ring with 1.

(e) Notice that in a ring without 1, not every element of a principal ideal (f) is a multiple of f. For instance, if $g = 8X + 4X^2 + 2X^3$, then $g = 2X + 2X + 2X + 2X + 2X(2X + X^2)$, so $g \in (2X)$, but g is not a multiple of $2X$ since $4 \notin R$.

(f) R is 1-dimensional and R is Noetherian since every ideal of R is an ideal of T and T is Noetherian.

(g) This example is from [Gilmer 1969c].

□ □ □ □ □

Example 151

Let R = (2) ⊆ Z. Let T = R[X].

(a) R and T are rings without 1.

(b) Every ideal of R is principal (this is trivial, since R is additively an infinite cyclic group and any subgroup of a cyclic group is cyclic).

(c) The prime ideals of R are (0) and $(2p)$, where p is a prime integer other than 2 or -2. Thus R is 1-dimensional.

(d) Not every ideal of R is a product of prime ideals. For instance, the ideal (18) is not. In a principal ideal domain (with 1), every ideal is a product of prime ideals.

(e) T is 2-dimensional, but not Noetherian. For instance, the ideal $(2X, 2X^2, 2X^3, \ldots)$ is not finitely generated. (Notice that $X \notin T$.)

(f) Obviously, instead of $R = (2)$, we could use (n), where n is any integer except $-1, 0$, or 1.

(g) Thus R is Noetherian and without 1, and $T = R[X]$ is not Noetherian. This is not possible if the ring of coefficients has 1.

(h) This example is from [Gilbert and Butts 1968].

Example 152

Let $R = K[X]/(X^3)$, where K is a field. Let x be the image of X and let T be the ideal (x) in R.

(a) T is a ring without 1. Every element of T is nilpotent of order 3 or less.

(b) If the field K is finite, then T is finite (as a vector space of dimension 2 over K) and hence Noetherian. If K is infinite, then T is not Noetherian, since the ideal $(a_1 X^2, a_2 X^2, a_3 X^2, \ldots)$ is not finitely generated, where $a_1, a_2, a_3, \ldots$ are distinct elements of K. (If $K = \mathbf{Q}$, we can use $a_i = 1/i$.)

(c) Let K be the field of p elements, for some prime integer p. Then the ideal $M = (x^2)$ is a maximal proper ideal of T, but is not a prime ideal since T/M is not a domain. In a ring with 1, every maximal proper ideal is prime.

Example 153

Let R be the subring of $\mathbf{Q}$ generated by the elements $2/3, 2/9, 2/27, 2/81, \ldots$.

(a) R is a ring without 1, consisting of all rational numbers m/n for which 2 divides m and n is a power of 3.

(b) R is not Noetherian, since $(2/3) \subset (2/9) \subset (2/27) \subset (2/81) \subset \ldots$ is an infinite properly ascending chain of ideals. Alternately, the ideal $(4/3, 4/9, 4/27, \ldots)$ is not finitely generated.

(c) Notice that any subring of $\mathbf{Q}$ which contains 1 is necessarily Noetherian.

(d) The only prime ideal of R is (0). Thus $\dim(R) = 0$.

(e) Let us consider $\mathbf{Q}$ with its usual absolute-value topology, and give R the subspace topology as a subset of $\mathbf{Q}$. Then R is a topological ring, and the completion of R contains 1, since $1 = 2/3 + 2/9 + 2/27 + \ldots$ in this topology.

This is not an I-adic topology for any ideal I of R.

(f) Instead of using 2 and 3 to construct R, we could use any other pair of prime integers to construct a similar example.

Example 154

Let R_0 be any quasi-local ring. Let M be its maximal ideal. Let $R = R_0[X]/(X^2, mX$ for all $m \in M)$. Let x be the image of X.

(a) R is integral over R_0, since it is generated over R_0 by 1 and the nilpotent element x. Therefore $\dim(R) = \dim(R_0)$.

(b) R is its own total quotient ring. That is, any element t of R which is not a unit is a zero-divisor. In fact, $tx = 0$ if t is not a unit.

(c) Thus one many have a ring of any dimension which is its own total quotient ring (a property which is automatic in dimension 0).

(d) If R_0 is Noetherian, so is R. R is trivially integrally closed in its own total quotient ring. Suppose R_0 is the ring of Example 28, Noetherian but not catenary. Then R is Noetherian and integrally closed in its own total quotient ring, but not catenary.

Example 155

Let K be a field. Let $R = K[X, Y, Z]/I$, where $I = (X, Y) \cap (Z)$. Let x, y, z be the images of X, Y, Z, respectively.

(a) R is a 2-dimensional Noetherian ring.

(b) Consider the principal ideal (x). By the principal ideal theorem, the prime ideals of R which are minimal over (x) may be of height 0 or 1. These include (x, y), which is of height 0, and (x, z), which is of height 1.

(c) Consider the ideal $J = (x, y + z)$, generated by two elements. The only prime ideal of R which contains J is the maximal ideal (x, y, z), which is of height 2.

(d) Thus in comparing (x) to J, we see that the number of generators increases by 1, while the height of the ideal increases by 2. The "height" of an ideal, in this sense, is the minimum value of ht(P) for all prime ideals P containing it.)

(e) This example is from [Eisenbud and Evans 1976].

□ □ □ □ □

Example 156

Let $R = K[[W, X, Y, Z]]/(WX - YZ)$, where K is a field.

(a) R is a 3-dimensional, integrally closed, complete, local Gorenstein domain.

(b) Let w, x, y, z be the images of W, X, Y, Z, respectively. Let $I = (w, y)$ and $J = (x, z)$. Then I and J are both prime ideals, each of height 1. (Since neither one is principal, this shows that R is not factorial and therefore not regular.)

(c) The ideal $I + J = M = (w, x, y, z)$ is the maximal ideal of R and therefore is of height 3.

(d) Thus ht(I) + ht(J) < ht($I + J$). This phenomenon could not occur in a regular local ring. (If T is a regular local ring, then it is a factorial domain and its prime ideals of height 1 are principal. If p, q are principal primes of T and $(p) \neq (q)$, then p, q is an R-sequence, and the ideal (p, q) is of grade 2. In a Cohen–Macaulay ring, the grade of an ideal and its height are the same.)

(e) This example is from [Eisenbud and Evans 1976].

□ □ □ □ □

Example 157

Let R be a non-Noetherian domain. (For instance, Example 1.) Let I be a non-finitely generated ideal of R. Let $T = R[X^3, X^4, X^5]$.

(a) Let $I_0 = (X^3, X^4, aX^5,$ for all $a \in I)$. Then I_0 is not finitely generated since I is not.

(b) $I_0^2 = (X^6, X^7, X^8)$ so, in particular, I_0^2 is finitely generated.

(c) Let $S = T[Y]$. Let S_0 be the subring $S_0 = T[aY$ for all $a \in I_0]$.

(d) In S_0, the ideal $P = (aY$ for all $a \in I_0)$ is not finitely generated and is prime.

(e) $P^2 = (X^6 Y^2, X^7 Y^2, X^8 Y^2)$ so, in particular, P^2 is finitely generated.

(f) Thus it is possible for an ideal (even a prime ideal) to be non-finitely generated while its square is finitely generated.

(g) This example is from [Gilmer 1972b].

Example 158

Let K be a field. Define a valuation v on $K(X, Y)$, taking on values in the additive group of real numbers, by defining v to be trivial on K, and $v(X) = 1, v(Y) = 2$. Let V be the valuation ring of v. Let M be the maximal ideal of V. Let n be an integer greater than 1 and let $t = X^{1/n}$. Let $R = V[t]$.

(a) The ideal $N = (t, M)$ is a maximal ideal of R.

(b) Notice that $t^i \in N^i$ for $i > 0$ and that $t^i \notin N^{i-1}$ for $i < n$. However, $t^n = X \in N^n$. Since $X \in M$ and $M = M^2$, N^n already contains $t^{n+1} = X^{1 + 1/n}$.

(c) It follows that $N^n = N^{n+1} = N^{n+2} = \ldots$. Thus the powers of N are all distinct until N^n is reached, after which they are all equal.

(d) This phenomenon cannot occur in a Prüfer domain. That is, if A is an ideal of a Prüfer domain and $A^m = A^{m+1}$ for some $m > 0$, then $A = A^2$. In this example, V is a 1-dimensional, non-Noetherian valuation domain and R is a simple integral extension of V, and not Prüfer.

(e) This example is from [Arnold and Gilmer 1967].

Example 159

Let K be a field with char$(K) = 0$. Let $L = K(X), L_1 = K(X^2)$, and $L_2 = K(X + X^2)$. Let $T = L[[Y]]$. Let M be the maximal ideal of T. Let $R_1 = L_1 + M, R_2 = L_2 + M$, and $R = K + M$.

(a) T is a Noetherian valuation domain. Since T is integral over either R_1 or R_2, and since L is a finite-dimensional extension of either L_1 or L_2, R_1 and R_2

are both 1-dimensional local domains.

(b) R is a 1-dimensional, quasi-local, non-Noetherian, integrally closed domain. Since $L_1 \cap L_2 = K$, we have $R = R_1 \cap R_2$.

(c) Thus we have two Noetherian domains, neither integrally closed, whose intersection is an integrally closed non-Noetherian domain. All of this takes place within the same quotient field, namely $L((Y))$.

(d) This example is from [Gilmer 1972a, p. 493, Exercise 18].

Example 160

Let K be a field. Let L be the field $L = K(Z, X + YZ)$ and let $L_0 = K(X, Y)$. Let $R = L_0[Z]$. Let $T = R \cap L$.

(a) R is a principal ideal domain with $K(X, Y, Z)$ as its quotient field.

(b) T is a Krull domain, since it is the intersection of a Krull domain with a subfield of its quotient field. Notice that $L \cap L_0 = K$. Actually we have $T = K[Z, X + YZ]$.

(c) Thus T is a 2-dimensional, regular, factorial domain. (Notice that Z and $X + YZ$ are algebraically independent over K.)

(d) Notice that $K(X, Y, Z)$ is a transcendental extension of L, and T is integrally closed in $K(X, Y, Z)$. In particular, R is not integral over T.

(e) This example is from [Gilmer 1972a, p. 504, Exercise 2].

Example 161

Let $K = \mathbf{Z}/(2)$ and let $L = \mathbf{Z}/(3)$. Let R be the group algebra of K (as an additive group) over K (as a field).

(a) R is a local ring whose group of units is isomorphic to the additive group K, i.e., the group of two elements.

(b) L is a field and therefore is trivially a local ring. The group of units of L is also isomorphic to K.

(c) Thus we have two nonisomorphic local rings with isomorphic groups of units.

(d) This example is due to Steve Landsburg.

□ □ □ □ □

Example 162

Let $R = K[X, Y]_{(X, Y)}$ and $T = K[X]_{(X)}$, where K is any field.

(a) R and T are both local factorial domains.

(b) R and T are clearly not isomorphic since $\dim(R) = 2$ and $\dim(T) = 1$.

(c) The units of T are polynomials in X with nonzero constant terms, and quotients of these. Each unit in T can be factored uniquely into a product of a nonzero element of K and powers of monic irreducible polynomials in X, possibly using negative exponents.

(d) It follows that the group of units of T is the direct sum of the group of units of K and a free abelian group. The generators of the free abelian group are the monic irreducible polynomials in X.

(e) Likewise, the group of units of R is also the direct sum of the group of units of K and a free abelian group. The generators of the free abelian group are the monic irreducible polynomials in X and Y.

(f) In fact, there are just as many monic irreducible polynomials in X and Y as there are in X alone; thus the groups of units of R and T are isomorphic.

(g) Obviously, similar remarks could be made about the group of units of any factorial domain.

(h) The groups of divisibility of R and T are not isomorphic.

(i) Let $S = R_{(X)}$. Then S and T are both Noetherian valuation domains, with isomorphic value groups and isomorphic groups of units, but S and T are not necessarily isomorphic. Specifically, if $K = \mathbf{Q}$, then S and T are not isomorphic because their respective maximal subfields $\mathbf{Q}(Y)$ and $\mathbf{Q}$ are not isomorphic.

□ □ □ □ □

Example 163

Let R be any ring which is not its own total quotient ring. Let $p \in R$ be neither a unit nor a zero-divisor. Let A be the free R-module on countably many generators x, y_1, y_2, y_3, ... Let B be the submodule generated by $x - py_1$, $x - p^2 y_2$, $x - p^3 y_3, ...$. Let $M = A/B$.

(a) Consider the submodules $(p)^n M$. Since $p^n y_i = x$ if $n = i$, $(p)^n M$ is generated by x and $p^n y_i$ for $n > i$.

(b) Therefore, $\cap (p)^n M$ = the submodule generated by x alone. Let $N = \cap (p)^n M$.

(c) It follows that $(p)N \neq N$, since p is not a unit. This phenomenon cannot occur for a finitely generated module over a Noetherian ring.

(d) To be more precise, if T is a Noetherian ring, I is an ideal of T, Q is a finitely generated T-module, and $Q_0 = \cap I^n Q$, then we have $IQ_0 = Q_0$. This is called the *Krull intersection theorem.*

(e) This example is from [Anderson 1975].

Example 164

Let R be a Dedekind domain which is not a principal ideal domain. Let K be the quotient field of R. Let $T = R + XK[[X]]$.

(a) T is a 2-dimensional Prüfer domain.

(b) Since R is a Dedekind domain, each of its ideals can be generated by only two elements. In T, as in most known examples of Prüfer domains, any finitely generated ideal can be generated by only two elements. (Only very recently have examples requiring more than two generators been produced.)

(c) In a Dedekind domain, in picking two suitable generators for an ideal, the first generator may be chosen at random from the nonzero elements of the ideal. (The choice of the second generator requires extreme care.) T does not have this property, which is sometimes called the *1½-generator property.*

Let $J_0 = (t, u)$ be a nonprincipal ideal of R. Let J be the ideal of T generated by t and u. Then $XK[[X]]$ is contained in J. If $m \in XK[[X]]$, there is no element a of T with $(a, m) = (t, u)$.

(d) It is possible to show that what has gone wrong is that $m \in XK[[X]]$, which is the Jacobson radical of T.

(e) This example is from [Gilmer and Heinzer 1970]. See also item #4 in the Further References section.

Example 165

Let $R = K[[X^2, X^3]][Y]$, where K is a field.

(a) R is a 2-dimensional Noetherian domain.

(b) Let $M = (X^2, X^3)$ and $I = (X^2 Y^2 - 1, X^3 Y - X^2)$. Notice that I is not a principal ideal and $IM = (XY - 1)M$.

(c) Let $f : I \oplus M \to R$ be the map $f(a, b) = a + b$. Since $f(1 - X^2 Y^2, X^2 Y^2) = 1$, this map is onto. Since R is projective as a module over itself, it can be shown that $I \oplus M$ must be isomorphic to $R \oplus$ (the kernel of f). Now the kernel of f consists of all elements $(a, -a)$ with $a \in I$. Now $I \cap M = IM = (XY - 1)M$ is isomorphic to M as an R-module. Therefore the kernel of f is isomorphic to M.

(d) Thus $I \oplus M$ is isomorphic to $R \oplus M$. Since I is not a principal ideal, it is not isomorphic to R as an R-module. This shows that if A, B, and C are modules (over some ring) and $A \oplus B$ is isomorphic to $A \oplus C$, we cannot conclude that B and C are isomorphic by cancelling the summand A.

(e) The ideal I is actually invertible. (See Example 134, which involves an almost identical ring and ideal.)

(f) This example is from [Vasconcelos 1967].

Example 166

Let p_1, p_2, p_3, ... be the sequence of positive prime integers. Let $R = \mathbf{Z}[X/p_1, X/p_2, X/p_3, \ldots]$.

(a) For $n > 0$, let $R_n = \mathbf{Z}[X/p_1 p_2 p_3 \ldots p_n]$. Each ring R_n is a 2-dimensional, regular, factorial domain. R is the union of the ascending sequence $R_1 \subseteq R_2 \subseteq R_3 \subseteq \ldots$, so R is 2-dimensional and integrally closed.

(b) R is not Noetherian, since the ideal generated by $X/p_1, X/p_2, X/p_3, \ldots$ is not finitely generated.

(c) Maximal ideals of R are of two sorts: those of the form (p, f), where p is a prime integer and the image of f is irreducible in $\mathbf{Z}/(p)[X/p]$, and those of the form $R \cap f\mathbf{Q}[X]$, where $f \in R$ and f is irreducible over $\mathbf{Q}$ (plus some conditions on f). Maximal ideals of the form $(X - k)$ for $k \in \mathbf{Z}$ are $(X + 1)$ and $(X - 1)$. It can be shown that each of these maximal ideals is finitely generated.

(d) If M is a maximal ideal of R, then R_M is isomorphic to a localization of either $\mathbf{Q}[X]$ or $\mathbf{Z}/(p)[X/p]$, where p is a prime integer. In either case, R_M is Noetherian.

(e) Thus R is not Noetherian or a polynomial ring over any subring of R, but it is locally Noetherian, locally a polynomial ring, and locally factorial.

(f) R is not a Krull domain. To see this, notice that the element X has infinitely many prime divisors, namely $(p_1), (p_2), (p_3), \ldots$.

(g) This example is from [Eakin and Silver 1972].

Example 167

Let $R = K[X_1, X_2, X_3, \ldots]$, where K is a field.

(a) Let I be the ideal $(X_1 X_2, X_3 X_4, \ldots, X_{2n}X_{2n+1}, \ldots)$. I is a radical ideal. (R/I has lots of zero-divisors but no nilpotent elements.)

(b) The prime ideals of R which are minimal over I are those of the form $P = (X_{1+d_1}, X_{3+d_3}, \ldots)$, where for each odd positive integer i, d_i is either 0 or 1. I is precisely the intersection of all such prime ideals P.

(c) However, any one given prime ideal P of this form is redundant in the intersection of all such.

(d) It is slightly interesting to notice that if P is a prime ideal of R minimal over I, then $\operatorname{ht}(P) = \dim(R/P) = \dim(R) = \infty$.

(e) This illustrates the pathological behavior of primary decomposition in an infinite-dimensional non-Noetherian ring.

(f) This example is from [Underwood 1969].

Example 168

Let K be a field. Let $L = K(X_1, X_2, X_3, \ldots)$ and for $m > 0$, let $L_m = K(X_1,$
$\ldots, X_m)$. Let n be a positive integer. Let V_m be an n-dimensional valuation do-
main of the form $V_m = L_m + M_m$, where M_m is the maximal ideal of V_m. Let
V be an n-dimensional valuation domain of the form $V = L + M$, where M is
the maximal ideal of V. Let $R = K + M$, and for $m > 0$, let $R_m = K + M_m$.

(a) R_m is an n-dimensional, integrally closed, quasi-local domain. R is also
an n-dimensional, integrally closed, quasi-local domain.

(b) A domain T is said to be of *valuative dimension n* (possibly infinite)
if n is the maximum of the Krull dimensions of all valuation domains lying
between T and its quotient field. This is denoted by $\dim_v(T) = n$.

(c) Notice that V lies between R and its quotient field. Therefore $\dim_v(R) >$
n. Since L is also contained in the quotient field of R, and there may be an infinite-dimensional valuation domain (or one of any finite dimension) lying
between K and L, actually $\dim_v(R) = \infty$. Likewise, V_m lies between R_m and
its quotient field, which must contain L_m, and there may be a valuation domain of any dimension $\leqslant n$ contained in L_m and containing K. Because of
this, $\dim_v(R_m) = m + n$.

(d) Suppose $m = n = 1$. Then one construction for V_1 is $V_1 = K(X_1) +$
$X_2 K(X_1)[X_2]_{(X_2)}$. Then $R_1 = K + M_1$, where $M_1 = X_2 K(X_1)[X_2]_{(X_2)}$.
We have $\dim_v(R_1) = 2$ since, for instance, the domain $T = K[X_1]_{(X_1)} + M_1$
is a 2-dimensional valuation domain contained in $K(X_1, X_2)$, which is the
quotient field of R_1.

(e) Notice that if we wish, we can arrange that the quotient field of R_m is
always L_{m+n}.

(f) This example is from [Gilmer 1972a, p. 372, Exercise 17].

Example 169

Let K be a field having $\mathrm{char}(K) = 0$. *Let R be the group ring of* **Q** *over K.*

(a) Elements of R can be thought of as polynomials in some indeterminates,
having coefficients in K and exponents in **Q**.

(b) R is clearly integral over $K[X, X^{-1}]$, which is a principal ideal domain,
so R is a 1-dimensional domain.

(c) In fact, if L is the quotient field of R, R is the integral closure of $K[X, X^{-1}]$ in L. It follows that R is a Prüfer domain.

(d) For each positive integer n, let G_n be the subgroup of $\mathbf{Q}$ generated by the element $1/n!$. Let R_n be the group ring of G_n over K. Since G_n is cyclic, for each n, G_n is isomorphic to $\mathbf{Z}$. Therefore, R_n is isomorphic to $K[X, X^{-1}]$, for each positive n.

(e) Since $\mathbf{Q}$ is the union of the ascending chain $G_1 \subseteq G_2 \subseteq G_3 \subseteq \ldots$ of subgroups, R is the union of the ascending chain $R_1 \subseteq R_2 \subseteq R_3 \subseteq \ldots$ of subrings. Since each ring R_n is a principal ideal domain, $R_n = K[X^{1/n!}]$, and thus Bézout, R is also a Bézout domain.

(f) R is actually almost Dedekind. Notice that R is not Noetherian, since $\mathbf{Q}$ is not finitely generated, so R is not Dedekind.

To prove that R is almost Dedekind is decidedly nontrivial. A 1-dimensional Prüfer domain is almost Dedekind if and only if it contains no *idempotent* maximal ideals, i.e., if M is a maximal ideal, then $M \neq M^2$.

(g) Let $\mathbf{Q}^+$ be the semigroup of nonnegative rational numbers. Let R^+ be the semigroup ring of $\mathbf{Q}^+$ over K. Then, just as with R, R^+ is 1-dimensional and Bézout; but is certainly not almost Dedekind, since the maximal ideal M generated by all X^a for $a \in \mathbf{Q}^+$ is definitely idempotent. R^+_M is a non-Noetherian valuation domain. R can be regarded as R^+ with this objection removed, i.e., $R = R^+[X^{-1}]$.

(h) Suppose $K = \mathbf{C}$. Let us, in this specialized case, prove that R is almost Dedekind.

Let P be a maximal ideal of R. Let $P_0 = P \cap \mathbf{C}[X, X^{-1}]$. For $n > 0$, let $P_n = P \cap R_n$. Then P is the union of the ascending chain $P_0 \subseteq P_1 \subseteq P_2 \subseteq \ldots$. Since, for each n, R_n is a principal ideal domain, P_n must be a principal ideal, say $P_n = (f_n)$. Since $\mathbf{C}$ is algebraically closed, f_n must be of the form $f_n = X^{1/n!} + b_n$, up to multiplication by a unit. Notice that $b_n^{n!} = b_0$.

Suppose $P = P^2$. Then $f_0 \in P$, so $f_0 \in P^2$. Now P^2 is the union of $P_0^2 \subseteq P_1^2 \subseteq P_2^2 \subseteq \ldots$, so $f_0 \in P_m^2$ for some $m > 0$. Since R_m is a principal ideal domain and $P_m = (f_m)$, $P_m^2 = (f_m^2)$, so f_m^2 must divide f_0 in R_m.

Since $\mathbf{C}$ is of characteristic 0, all of the field extensions involved are separable. This makes it absurd for f_m^2 to divide f_0. (Another way to look at it is that $(X^{1/m!} + b_m)^2$ must divide $(X^{1/n!})^{n!} + b_n^{n!}$, or if we relabel everything, $(y + z)^2$ must divide $y^r + z^r$, inside a ring of polynomials in characteristic 0.)

Therefore $P \neq P^2$ and R must be almost Dedekind.

(i) One obvious question about the ring in (h) is: Since R_P is a Noetherian valuation domain, its maximal ideal is principal; what element generates the maximal ideal? The answer is that any of the elements f_n will do.

(j) If K is some other field of characteristic 0, the above proof can be carried through, although we might have to deal with nonlinear irreducible polynomials. If char$(K) \neq 0$, inseparability problems arise and the entire situation becomes quite complex.

(k) This example is from [Gilmer and Parker 1974].

□ □ □ □ □

Example 170

Let $R_0 = K[[X]]$, where K is a field. Let $R = R_0[Y]$. Let $M = (X, Y)$ and $N = (X, Y + 1)$. M and N are maximal ideals of R. Let $S = R - (M \cup N)$.

(a) R_0 is a complete, Noetherian, valuation domain. R is a 2-dimensional, regular, factorial domain.

(b) R_S is a 2-dimensional, regular, factorial domain having precisely two maximal ideals, namely M_S and N_S.

(c) The only prime ideals of R contained in $M \cap N$ are (0) and (X). (To see this, suppose P' is such a prime ideal and consider $P' \cap R_0$. It is to rule out the possibility of $P' \cap R_0 = (0)$ and $P' \neq (0)$ that the completeness of R_0 is needed.)

(d) Therefore, the only prime ideals of R_S which are contained in $M_S \cap N_S$ are (0) and (X).

(e) Let L be the quotient field of R_S. Let $b = Y(Y + 1)$. Then b is in $M_S \cap N_S$, but not in (X), and no prime ideal of R_S of height 1 contains both X and b. Let $Q = (Z - b/X)L[Z] \cap R_S[Z]$. Then Q is a prime ideal of $R_S[Z]$ of height 1 and $XZ - b$ is in Q.

(f) In fact, Q is the only prime ideal of $R_S[Z]$ of height 1 which contains $XZ - b$. (To see this, suppose Q' is such a prime ideal and consider $Q' \cap R_S$.)

(g) Notice that $X \in (X) \subseteq M_S \cap N_S$ and that $b \in M_S \cap N_S$. Let us denote $M_S R_S[Z]$ and $N_S R_S[Z]$ by M' and N', respectively.

(h) Notice that Q is contained in $M' \cap N'$, so Q is contained in $(M', Z) \cap (N', Z)$. In fact, if P is a prime ideal of $R_S[Z]$ and $Q \subseteq P \subseteq (M', Z) \cap (N', Z)$, then $Q = P$. This can be proven by examining $P \cap R_S$, which must be either (0) or (X). If $P \cap R_S = (X)$, then $X \in P$. Since $Q \subseteq P$, we have $XZ - b \in P$. Therefore $b \in P$ so $b \in P \cap R_S = (X)$, which contradicts the earlier statement that no prime ideal of R_S of height 1 can contain both b and X. If $P \cap R_S = (0)$, then $(R_S[Z])_P$ is a localization of $L[Z]$, and so is 1-dimensional. Yet $Q \subseteq$

P; therefore, $Q = P$.

(i) Let $T = R_S[Z]/Q$. T is a Noetherian domain with $\dim(T) = 2$. The images of (M', Z) and (N', Z) are each maximal ideals of T, of height 2. From (h), it is clear that the intersection of these two maximal ideals does not contain any prime ideal of T of height 1.

(j) Notice that T is actually Cohen–Macaulay. I don't know if it is regular or factorial.

(k) This example is from [McAdam 1976].

□ □ □ □ □

Example 171

Let $R_0 = K[[X]]$, where K is a field of characteristic $p > 0$. Let R be the sub-ring $R = K[[X^2, X^3]]$.

(a) R_0 is a Noetherian valuation domain. R_0 therefore has group of divisibility $G(R_0) \cong \mathbf{Z}$.

(b) R is a 1-dimensional local domain. Its integral closure is R_0.

(c) Since R_0 is an overring of R, $\mathbf{Z}$ must be a homomorphic image of $G(R)$. What is $G(R)$? Consider the elements $a + X$, which are in R_0, and therefore in the quotient field of R. We have $(a + X)^p = a^p + X^p$, which is in R and is a unit. Therefore, $G(R)$ has some elements of finite order, and is not torsion-free.

(d) In fact, it can be shown that $G(R)$ is isomorphic to $\mathbf{Z} \oplus K^{(+)}$, where $K^{(+)}$ is the additive group of the field K. In $G(R)$, an element (n, b) is positive if and only if $n \geqslant 2$.

(e) We could use an appropriately localized polynomial ring instead of a power series ring, if we wished, to obtain this example of a ring which is not root-closed.

(f) This example is from [Gilmer 1972a, pp. 186–187, Exercise 19].

□ □ □ □ □

Example 172

Let L be the algebraic closure of **Q**. *Let K be the subfield of L consisting of all elements whose minimal polynomials over* **Q** *can be solved by radicals. Let R =* $K + XL[X]$.

(a) R is a 1-dimensional quasi-local domain.

(b) K is not algebraically closed in L since there are polynomials (e.g., of degree 5) which cannot be solved by radicals over **Q**. Thus R is not integrally closed.

(c) R does contain each element t of its quotient field $L(X)$ for which some power t^n is in R. Therefore, the group of divisibility of R is torsion-free, even though R is not integrally closed. (This illustrates that root-closure is strictly weaker than integral closure.)

(d) This example is from [Gilmer 1972a, p. 184, Exercise 6].

□ □ □ □ □

Example 173

Let $R = K[X, Y]/(X^2, XY, Y^2)$, *where* $K = $ **Z**$/(2)$. *Let* x, y *be the images of X, Y, respectively, so* $R = K[x, y]$.

(a) R is a 0-dimensional local ring.

(b) Let A be a 3-dimensional vector space over K. Let a, b, c be a basis for A. Make A into an R-module by defining $xa = yb = xc = yc = 0$ and $xb = ya = c$. Notice that $(x + y)(a + b) = 0$. A is a *faithful* R-module; i.e., there is no nonzero element $r \in R$ such that $rA = 0$. However, each element of A has a nonzero annihilator.

(c) This example was relayed to me by I. Kaplansky.

□ □ □ □ □

Example 174

Let $R_0 = $ **C**$[X]$. *Let S be the multiplicatively closed set generated by* $X - a$ *for* $a \in $ **C**, $a \notin $ **Z**. *Let* $R = (R_0)_S$.

(a) R is a 1-dimensional, Hilbert, principal ideal domain having countably many maximal ideals, all of which are of the form $(X - i)$ for some $i \in \mathbf{Z}$.

(b) Notice that for each maximal ideal $(X - i)$ we have $R/(X - i) \cong \mathbf{C}$.

(c) Now let $T = R[Y_i$ for all $i \in \mathbf{Z}]$. T is a Hilbert ring, as a ring of polynomials over R.

(d) In T, let $M = (Y_i(X - i) - 1$ for all $i \in \mathbf{Z})$. M is a maximal ideal, so that T/M is a field.

(e) Specifically, T/M is isomorphic to $\mathbf{C}(X)$. However, $M \cap R = (0)$. Thus R is Hilbert, T is a (large) polynomial extension of R, and M is a maximal ideal of T, but $M \cap R$ is not a maximal ideal in R.

(f) This would be impossible if T were a polynomial ring in only finitely many variables.

(g) This example is from [Gilmer 1971].

□ □ □ □ □

Example 175

Let $K = K_0(X_1, X_2, X_3, \ldots)$, where K_0 is any field. Let V_1 be a 1-dimensional valuation domain with quotient field K, of the form $V_1 = K_0 + M_1$, where M_1 is the maximal ideal of V_1. (One way to construct V_1 is to define a real-valued valuation v on K by choosing the numbers $v(X_i)$ to be algebraically independent transcendental numbers and requiring v to be trivial on K_0.) For each integer $n \geqslant 2$, let L_n be the field $L_n = K(X_1, \ldots, X_n, X_{2n}, X_{2n+1}, \ldots)$. Thus $K = L_n(X_{n+1}, \ldots, X_{2n-1})$, and tr. deg.$(K/L_n) = n - 1$.

Now $V_1 \cap L_n$ is a valuation domain with quotient field L_n. We can find a valuation domain V_n with quotient field K, satisfying the following conditions:

(i) $V_n \cap L_n = V_1 \cap L_n$,

(ii) $\dim(V_n) = n$,

(iii) $V_n = K_0 + M_n$, where M_n is the maximal ideal of V_n.

(For instance, to construct V_2, we can consider $S_2 = L_2[X_3]$, and let N be the maximal ideal of $(S_2)_{(X_3)}$. Then $V_2 = V_1 + N$ satisfies the above conditions.)

Now let $P =$ the intersection of all the M_i and let $R = K_0 + P$. Let T be the integral closure of R.

(a) R is a quasi-local domain with maximal ideal P. R is infinite-dimensional.

(b) It can be shown that T is actually a Prüfer domain with quotient field

K, and that T = the intersection of all the valuation domains V_i. Since $\dim(R) = \infty$, T is also infinite-dimensional.

(c) It can be shown that any valuation domain lying between T and K is contained in some one of the V_i; thus any such valuation domain is finite-dimensional.

(d) Thus T is integral over R, and R has a single maximal ideal, of infinite height, while the maximal ideals of T are all of finite height. This would be impossible if R were finite-dimensional.

(e) I don't know whether R is catenary or T is Bézout.

(f) This example is due to [Heinzer 1971].

□ □ □ □ □

Example 176

Suppose F_1, F_2, F_3, ... are fields. Let R = the direct product of the F_i.

(a) R is a 0-dimensional non-Noetherian ring with no nonzero nilpotent elements.

(b) Thus R is von Neumann regular and every R-module must be flat.

(c) In particular, if I is a non-finitely generated ideal of R, e.g., $I = (0) \times F_2 \times F_3 \times \ldots$, then R/I is flat over R, and so is $R[X]/(I, X)$.

(d) One reason this is of interest is that if T is a domain and $T[X]/J$ is flat over T, then J must be finitely generated.

(e) R is a rather typical example of a (commutative) von Neumann regular ring; any such ring can be embedded into a suitable direct product of fields.

□ □ □ □ □

Example 177

Let D be a countable almost Dedekind domain which is not Dedekind. Let M be a maximal ideal of D which is not finitely generated, so that $M = (m_1, m_2, m_3, \ldots)$. Let $T = D[[X]]$.

(a) One way to produce D is to use the ring of Example 59.

(b) In T there are (at least) several prime ideals which lie over M in D. One of these is $M^* = M[[X]]$, which contains all power series in X all of whose coefficients are from M. Another is $MD[[X]]$, which is smaller, containing all linear combinations $a_1 t_1 + \ldots + a_n t_n$, where the a_i are from M and the t_i are from T.

(c) Consider the element $f = m_1 X + m_2 X^2 + m_3 X^3 + \ldots$. Clearly, $f \in M^*$ and $f \notin MD[[X]]$. Thus M^* is definitely of height > 1.

(d) In fact, $\dim(T) = \infty$ and $\dim(T/M^*) = 1$.

(e) Notice that D_M is a Noetherian valuation domain. On the other hand, T_{M^*} is not a valuation domain.

(f) It can be shown that for a domain D and a prime ideal P, if $D[[X]]_{P^*}$ is a valuation domain, where $P^* = P[[X]]$, then D_P is also a valuation domain. If $PD[[X]] = P^*$, then $D[[X]]_{P^*}$ is a valuation domain.

(g) This example is from [Arnold and Brewer 1973] and [Arnold 1973].

Example 178

Let $R_0 = \mathbf{Q}[X, Y]$. Let $M = (X, Y)$. Let $T = (R_0)_M$.

(a) R_0 is a 2-dimensional regular domain and T is a 2-dimensional, regular, local domain.

(b) Since $\mathbf{Q}$ is a countable field, R_0 and T are countable. Since T is countable, it has only countably many principal prime elements, and therefore only countably many prime ideals of height 1. (Of course, T is factorial, since it is a regular local domain.)

(c) Let $P_1, P_2, P_3, \ldots$ be the height-1 prime ideals of T. For each positive integer n, let $S_n = P_1 \cup P_2 \cup \ldots \cup P_n$ and let $T_n = T_{S_n}$.

(d) Suppose $N = M_M$ is the maximal ideal of T. Then $N =$ the union of all the P_n. Thus each ring T_n is of dimension 1, and we have $T_1 \supset T_2 \supset T_3 \supset \ldots$. The prime ideals of each ring T_n are precisely those corresponding to $P_1, \ldots, P_n$.

(e) The intersection of all the T_n is T itself. Thus the intersection of a descending chain of 1-dimensional rings may be of dimension 2.

(e) Clearly, a similar construction could be devised to produce a descending chain of rings of dimension n whose intersection is a ring of dimension $n + k$, for given positive integers n and k.

(f) This example was suggested by K. Goodearl.

Example 179

Let K be a field. Let $\mathbf{Z}^+$ be the set of all nonnegative integers. Let R be the set of all eventually constant functions $f : \mathbf{Z}^+ \to K$, i.e., functions f such that there is some $N \in \mathbf{Z}^+$ with $f(N) = f(N + 1) = f(N + 2) = \ldots$. R is a ring under pointwise operations.

(a) R is von Neumann regular. Why? If $f \in R$, we can define $g : \mathbf{Z}^+ \to K$ by $g(n) = 0$ if $f(n) = 0$, and $g(n) = (f(n))^{-1}$ if $f(n) \neq 0$. Since f is eventually constant, so is g, and $g \in R$. Then $fgf = f$.

(b) Since R is von Neumann regular, it is 0-dimensional. Let us examine its maximal ideals. For each $n \in \mathbf{Z}^+$, there is a maximal ideal M_n consisting of all $f \in R$ with $f(n) = 0$. In addition, there is one more maximal ideal M^*, consisting of all $f \in R$ which are eventually 0. These are all of the maximal ideals of R. Notice that for each $n \in \mathbf{Z}^+$, M^* contains $M_n \cap M_{n+1} \cap M_{n+2} \cap \ldots$.

(c) An additive group whose additive structure resembles that of R was used in Example 60.

(d) Notice that R is von Neumann regular and is not isomorphic to a direct sum of fields, unlike Example 176.

Example 180

Let S be any topological space. Let R be the set of all continuous functions $S \to \mathbf{R}$, where $\mathbf{R}$ is given the usual topology. With the usual pointwise operations, R is a ring, often denoted by $C(S)$.

(a) Notice that R always exists; in particular, constant functions are always continuous. Thus R contains an isomorphic copy of $\mathbf{R}$.

(b) The topological structure of S and the ring structure of R are intimately related. Here are some instances:

S is connected, i.e., is not a disjoint union of any two proper open subsets, if and only if R is not a nontrivial direct sum of any two rings.

If S is completely regular and contains more than two points, then R is not a domain.

R is von Neumann regular if and only if S is a *P-space*, i.e., if and only if every countable intersection of open sets is open and S is completely regular.

(c) In some cases, S can be recovered from R. More precisely, if S and S'

are two compact Hausdorff spaces, then $C(S)$ and $C(S')$ are isomorphic if and only if S and S' are homeomorphic.

(d) On the other hand, if S and S' are two sets, having different numbers of elements, and given the indiscrete topologies, then $C(S)$ and $C(S')$ are both isomorphic to $\mathbf{R}$, since the only continuous functions on an indiscrete space are the constants.

(e) In any case, regardless of S, some statements about the ring structure of R hold; for instance, each prime ideal of R is contained in a unique maximal ideal.

(f) Also see Example 133.

(g) The rings $C(S)$ have been studied in great detail. See [Gillman and Jerison 1960]. For material on point-set topology, see [Dugundji 1966].

Bibliography

D. D. Anderson, 1975. The Krull intersection theorem, Pacific J. Math. 57, 11–14.

J. Arnold, 1973. Krull dimension in power series rings, Trans. Amer. Math. Soc. 177, 299–304.

J. Arnold and J. Brewer, 1973. When $D[[X]]_{P[[X]]}$ is a valuation ring, Proc. Amer. Math. Soc. 37, 326–332.

J. Arnold and R. Gilmer, 1967. Idempotent ideals and unions of nets of Prüfer domains, J. Sci. Hiroshima Univ., Ser. A-1, 31, 131–145.

J. Arnold and R. Gilmer, 1974a. The dimension sequence of a commutative ring, Amer. J. Math. 96, 385–408.

J. Arnold and R. Gilmer, 1974b. Dimension theory of commutative rings without identity, J. Pure and Applied Algebra 5, 209–231.

M.-J. Bertin, 1967. Anneaux d'invariants d'anneaux de polynomes, en caracteristique p, C. R. Acad. Sci. Paris 264, 653–656.

N. Bourbaki, 1972. *Commutative Algebra,* Paris, Hermann, translated by the author.

H. S. Butts and R. C. Phillips, 1965. Almost multiplication rings, Canad. J. Math. 17, 265–277.

C. L. Chuang and P. H. Lee, 1977. Noetherian rings with involution, Chinese J. Math. 5, 15–19.

L. Claborn, 1966. Every abelian group is a class group, Pacific J. Math. 18, 219–222.

J. Cohen, 1969. A note on homological dimension, J. Algebra 11, 483–487.

H. Cohn, 1962. *A Second Course on Number Theory,* New York, Wiley.

V. I. Danilov, 1970. On a conjecture of Samuel, Math. USSR–Sb. 12, 368–386.

J. David, 1972. A non-Noetherian factorial ring, Trans. Amer. Math. Soc. 169, 495–502.

J. Dieudonne, 1967. *Topics in Local Algebra,* Notre Dame, Notre Dame Press.

J. Dugundji, 1966. *Topology,* Boston, Allyn and Bacon.

P. Eakin, 1968. The converse to a well-known theorem on Noetherian rings, Math. Ann. 177, 278–282.

P. Eakin and W. Heinzer, 1968. Some open questions on minimal primes of Krull domains, Canad. J. Math. 20, 1261–1264.

P. Eakin and W. Heinzer, 1970. Non-finiteness in finite-dimensional Krull domains, J. Algebra 14, 333–340.

P. Eakin and W. Heinzer, 1973. More noneuclidean PIDs and Dedekind domains with prescribed class group, Proc. Amer. Math. Soc. 40, 66–68.

P. Eakin and K. Kubota, 1972. A note on the uniqueness of ring coefficients in polynomial rings, Proc. Amer. Math. Soc. 32, 333–341.

P. Eakin and J. Silver, 1972. Rings which are almost polynomial rings, Trans. Amer. Math. Soc. 174, 425–449.

D. Eisenbud and E. Evans, 1976. A generalized principal ideal theorem, Nagoya Math. J. 62, 41–53.

D. Fields, 1969. Zero divisors and nilpotent elements in power series rings, Proc. Amer. Math. Soc. 27, 427–433.

R. Fossum, 1973. *The Divisor Class Group of a Krull Domain,* New York, Springer-Verlag.

R. Fossum and P. Griffith, 1975. A complete local factorial ring of dimension 4 which is not Cohen–Macaulay, Bull. Amer. Math. Soc. 81, 111–113.

J. Gilbert and H. S. Butts, 1968. Rings satisfying the 3 Noether axioms, J. Sci. Hiroshima Univ., Ser. A-1, 32, 211–224.

L. Gillman and M. Jerison, 1960. *Rings of Continuous Functions,* Princeton, Van Nostrand.

R. Gilmer, 1967a. If $R[X]$ is Noetherian, R contains an identity, Amer. Math. Monthly 74, 700.

R. Gilmer, 1967b. A note on the quotient field of $D[[X]]$, Proc. Amer. Math. Soc. 18, 1138–1140.

R. Gilmer, 1969a. A note on generating sets for invertible ideals, Proc. Amer. Math. Soc. 22, 426–427.

R. Gilmer, 1969b. Integral dependence in power series, J. Algebra 11, 488–502.

R. Gilmer, 1969c. Commutative rings in which each prime ideal is principal, Math. Ann. 183, 151–158.

R. Gilmer, 1971. On polynomial rings over a Hilbert ring, Mich. Math. J. 18, 205–212.

R. Gilmer, 1972a. *Multiplicative Ideal Theory,* New York, Marcel Dekker.

R. Gilmer, 1972b. On factorization into prime ideals, Comment. Math. Helv. 47, 70–74.

R. Gilmer, 1973. Prüfer-like conditions on the set of overrings of an integral domain, in *Conference on Commutative Algebra,* New York, Springer-Verlag.

R. Gilmer, 1975. On polynomial and power series rings over a commutative ring, Rocky Mountain J. Math. 5, 157–175.

R. Gilmer and W. Heinzer, 1966. On the complete integral closure of an integral domain, J. Austral. Math. Soc. 6, 351–361.

R. Gilmer and W. Heinzer, 1970. On the number of generators of an invertible ideal, J. Algebra 14, 139–151.

R. Gilmer and T. Parker, 1974. Semigroup rings are Prüfer rings, Duke Math. J. 41, 219–230.

E. Hamann, 1975a. On the R-invariance of $R[X]$, J. Algebra 35, 1–16.

E. Hamann, 1975b. On power-invariance, Pacific J. Math. 61, 153–159.

W. Heinzer, 1969. Some remarks on complete integral closure, J. Austral. Math. Soc. 9, 310–314.

W. Heinzer, 1971. Integral ring extensions and prime ideals of infinite rank, Proc. Amer. Math. Soc. 28, 344–346.

W. Heinzer and J. Ohm, 1971. Locally Noetherian commutative rings, Trans. Amer. Math. Soc. 158, 273–284.

M. Hochster, 1972. Nonuniqueness of coefficient rings in a polynomial ring, Proc. Amer. Math. Soc. 34, 81–82.

M. Hochster, 1974. Grade-sensitive modules and perfect modules, Proc. London Math. Soc. 29, 55–76.

P. Jaffard, 1956. Un contre-exemple concernant les groupes de divisibilite, C. R. Acad. Sci. Paris 243, 1264–1268.

I. Kaplansky, 1974a. *Fields and Rings,* second edition, Chicago, Univ. of Chicago Press.

I. Kaplansky, 1974b. *Commutative Rings,* revised edition, Chicago, Univ. of Chicago Press.

I. Kaplansky, 1976. *Topics in Commutative Ring Theory,* Chicago, Univ. of Chicago Math. Dept.

M. Larsen and P. McCarthy, 1971. *Multiplicative Theory of Ideals,* New York, Academic Press.

H. Matsumura, 1970. *Commutative Algebra,* New York, Benjamin.

S. McAdam, 1976. A Noetherian example, Comm. in Algebra 4, 245–247.

J. Mott, 1973. The group of divisibility and its applications, in *Conference on Commutative Algebra,* New York, Springer-Verlag.

J. Mott, 1974. Convex directed subgroups of a group of divisibility, Canad. J. Math. 26, 532–542.

M. P. Murthy, 1976. *Commutative Algebra,* vols. 1 and 2, Chicago, Univ. of Chicago Math. Dept.

K. R. Nagarajan, 1968. Groups acting on Noetherian rings, Nieuw. Arch. Wisk. 72, 25–29.

M. Nagata, 1962. *Local Rings,* New York, Interscience.

M. Nagata, 1970. *Field Theory,* New York, Marcel Dekker.

M. Nakano, 1953. Idealtheorie in einem speziellen unendlichen algebraischen Zahlkorper, J. Sci. Hiroshima Univ., Ser. A-1, 16, 425–439.

J. Nichols, 1972. Remarks on homological dimension, Rocky Mountain J. Math. 2, 25—29.

D. Northcott, 1953. *Ideal Theory,* Cambridge, Cambridge Univ. Press.

B. Osofsky, 1968. Global dimension of commutative rings with linearly ordered ideals, J. London Math. Soc. 44, 183—185.

P. Quartararo and H. Butts, 1975. Finite unions of ideals and modules, Proc. Amer. Math. Soc. 52, 91—96.

J. Rotman, 1970. *Notes on Homological Algebra,* New York, Van Nostrand.

P. Salmon, 1966. Su in problema posto da P. Samuel, Atti Acad. Naz. Lincei. Rend. Cl. Sci. Fis. Mat. Natur. 40, 801—803.

P. Samuel, 1969. Unique factorization, Amer. Math. Monthly 75, 945—952.

P. Sheldon, 1974. Prime ideals in GCD domains, Canad. J. Math. 26, 98—107.

D. H. Underwood, 1969. On some uniqueness questions in primary representations of ideals, J. Math. Kyoto Univ. 9-1, 69—94.

W. Vasconcelos, 1967. Ideals and cancellation, Math. Z. 102, 353—355.

W. Vasconcelos, 1969. On finitely generated flat modules, Trans. Amer. Math. Soc. 138, 505—512.

W. Vasconcelos, 1970. Simple flat extensions, J. Algebra 16, 105—107.

B. L. van der Waerden, 1970. *Algebra,* vol. 1, seventh edition, vol. 2, fifth edition, New York, Frederick Ungar, translated by F. Blum and J. R. Schulenberger.

W. C. Waterhouse, 1971. Divisor classes in pseudo-Galois extensions, Pacific J. Math. 36, 541—548.

O. Zariski and P. Samuel, 1958. *Commutative Algebra,* vol. 1, New York, Van Nostrand.

Further References

In the search for examples of commutative rings, many were encountered which have not been included in this collection, for various reasons; some were too obscure, too specialized, or too complex. In addition, there are some theorems which establish the existence of examples without providing an explicit method of construction.

The following is an admittedly incomplete list of papers which contain examples of commutative rings. In some cases, the papers were written to announce examples; in other cases, the examples are incidental. In addition, there are many examples, interesting and otherwise, to be found in [Bourbaki 1972], [Gilmer 1972a], and [Fossum 1973], and in many of the papers and books listed in the Bibliography.

#1   T. Kabele, 1971. Regularity conditions in non-Noetherian rings, Trans. Amer. Math. Soc. 155, 363–374.

This paper contains, among other things, an example of a quasi-local ring whose maximal ideal M is finitely generated and has pd(M) $< \infty$, but M is not generated by an R-sequence. There is also an example of a quasi-local domain in which a permutation of an R-sequence may fail to be an R-sequence.

In Noetherian rings, neither of these situations is possible. In non-Noetherian rings, the various definitions of regularity fail in various ways; conditions which are equivalent for Noetherian rings may fail to be equivalent for non-Noetherian rings. For instance, if R is a 1-dimensional non-Noetherian valuation domain with maximal ideal M, then $M^2 = M$, so M/M^2 is of dimension 0 as a vector space over R/M.

#2   T. Ogoma, to appear. Non-catenary pseudo-geometric normal rings, Japan J. of Math.

R. Heitmann, to appear. A non-catenary normal domain.

For many years after the construction by Nagata of a noncatenary Noetherian domain (Examples 28 and 29), the question of whether such a ring could also be integrally closed remained open. Very recently—so recently that at the time of this writing, neither of these papers has yet been published—Ogoma has constructed a domain which is three-dimensional, integrally closed, and local. Heitmann's paper provides an alternate construction of Ogoma's example.

#3   H. -W. Schulting, to appear. Uber die Erzeugendenanzahl invertierbarer
     Ideals in Prüferringen.

For some years it was conjectured that in any Prüfer domain, any finitely generated ideal could actually be generated by only two generators. Various weaker statements have been proven (see #4) but the main conjecture has only recently been disproven. In this paper, Schulting presents an example of a 2-dimensional Prüfer domain containing a finitely generated ideal that requires three generators. The ring in question is a subring of $R(X, Y)$.

#4   R. Heitmann and L. Levy, 1975. 1½ and 2 generator ideals in Prüfer do-
     mains, Rocky Mountain J. Math. 5, 361–373.

In any Dedekind domain, any ideal can be generated by at most two generators, and in fact, a stronger statement is true: one of the two may be chosen at random from among the nonzero elements of the ideal (while the second must be chosen with great care). These facts have given rise to the conjecture mentioned it the preceding item, and to the following question: For which Prüfer domains is it true that any finitely generated ideal I can be generated by at most two elements, one of which may be chosen at random from $I - IJ$, where J is the Jacobson radical of the ring? (This is the 1½ generator property. See also Example 164.)

In this paper, it is proven that any 1-dimensional Prüfer domain has the 1½ generator property (which corroborates the theorem on Dedekind domains, as any Dedekind domain with a nontrivial Jacobson radical is semi-local and therefore a principal ideal domain).

There is also an example of a 2-dimensional Prüfer domain and an ideal which requires two "honest" generators. Thus, not all Prüfer domains, even in dimension 2, have the 1½ generator property.

#5   P. Hill, 1972. On the complete integral closure of a domain, Proc. Amer.
     Math. Soc. 36, 26–30.

     D. Lantz, 1975. Finite Krull dimension, complete integral closure and
     GCD-domains, Comm. Alg. 3, 951–958.

Let R be a domain. Let R' denote the complete integral closure of R, and define $R^{(n)}$ inductively by $R^{(n)} = (R^{(n-1)})'$ for $n > 1$. It is well known that we may have $R' \neq R^{(2)}$; see Examples 91 and 92, for instance.

In Hill's paper is a construction for a family of Bézout domains T_n which are such that $T_n' \subset T_n^{(2)} \subset T_n^{(3)} \subset \ldots \subset T_n^{(n)} = T_n^{(n+1)} = \ldots$. There is also a Bézout domain S for which the overrings $S^{(n)}$ are all distinct (for all positive integers n).

Lantz provides a stronger construction; he constructs, for each countable ordinal σ, a 2-dimensional Bézout domain D_σ such that $D_\sigma^{(\sigma)} \neq D_\sigma^{(\sigma+1)} = D_\sigma^{(\sigma+2)}$. He also proves that $D^{(\Omega)}$ is completely integrally closed for any domain D, where Ω denotes the first uncountable ordinal.

All of these constructions are done by means of constructing suitable lattice-ordered abelian groups to be the groups of divisibility, and then using the construction given in Chapter 2, Section B (or an equivalent construction).

#6   J. David, A characteristic-zero non-Noetherian factorial ring of dimension three, Trans. Amer. Math. Soc. 180, 315–325.

      R. Gilmer, 1974. A two-dimensional non-Noetherian factorial ring, Proc. Amer. Math. Soc. 44, 25–30.

      J. Brewer, D. Costa, and E. Lady, 1975, Prime ideals and localization in commutative group rings, J. Alg. 34, 300–308.

These three papers, as well as [David 1972], give examples of non-Noetherian factorial domains, in various characteristics and dimensions. For some years there was considerable speculation that factorial domains of low Krull dimension were necessarily Noetherian. (Of course, Example 1 demonstrates that an infinite-dimensional factorial domain need not be Noetherian.) (Also see Example 80.)

#7   K. Fujita, 1977. Three-dimensional unique factorization domain which is not catenary, J. Alg. 49, 411–414.

Even after the construction of non-Noetherian factorial domains (see the preceding items), a similar conjecture remained open for some time. Specifically, examples were sought of noncatenary factorial domains of low Krull dimension. (Again, Example 1 demonstrates that, in general, factorial domains need not be catenary.) It was proven that a 3-dimensional, Noetherian , factorial domain was also catenary.

This paper presents an example (as the title states) of a 3-dimensional factorial domain which is not catenary, and therefore necessarily not Noetherian.

Since any 2-dimensional domain is catenary, this is as good an example as one could want.

#8   M. Hochster, 1969. Prime ideal structure in commutative rings, Trans. Amer. Math. Soc. 142, 43–60.

W. Lewis, 1973. The spectrum of a ring as a partially ordered set, J. Alg. 25, 419–434.

The *spectrum* of a commutative ring R, denoted $\mathrm{Spec}(R)$, is the set of its prime ideals. This set is partially ordered by inclusion and usually is viewed as a topological space, with a topology derived from the partial order. The spectra of commutative rings are of tremendous interest in algebraic geometry.

It is reasonable to ask which partially ordered sets are eligible to be $\mathrm{Spec}(R)$ for some commutative ring R. Hochster's paper provides a tremendous amount of information on this matter; for instance, he shows that any finite partially ordered set is eligible. Lewis provides an explicit construction, by means of induction, of a ring having any given finite partially ordered set as its spectrum.

#9   C. P. Leedham-Green, 1972. The class group of Dedekind domains, Trans. Amer. Math. Soc. 163, 493–500.

This paper provides an alternate proof of Claborn's theorem, which states that for any abelian group A, there is a Dedekind domain D for which $A \cong \mathrm{Cl}(D)$. Specifically, it is shown that D may always be constructed as a quadratic extension of a principal ideal domain.

See also Example 50, and also [Eakin and Heinzer 1973]. In the latter paper, a construction is given by which any finitely generated abelian group is isomorphic to $\mathrm{Cl}(R)$ for some domain R, lying between $\mathbf{Z}[X]$ and $\mathbf{Q}(X)$.

Claborn's construction, of course, is given in [Claborn 1966].

#10 P. Eakin, 1972. A note on finite-dimensional subrings of polynomial rings, Proc. Amer. Math. Soc. 31, 75–80.

By definition, any homomorphic image of a polynomial ring over a field is an affine ring. It is reasonable to ask which subrings of a polynomial ring are affine. It is easy to see that finitely generated subrings are all affine; thus the interesting subrings are all infinitely generated.

In this paper, it is shown that if K is a field, then any 1-dimensional subring R of $K[X, Y]$ is affine, and if R is integrally closed, then $R = K[f]$ for some $f \in K[X, Y]$. There are also some partial results in the 2-dimensional case; for instance, an example is given of a subring of $K[X, Y]$ which is a Noetherian Krull domain, but is not affine. (See also Example 3.)

#11 W. Hodges, 1974. Six impossible rings, J. Alg. 31, 218–244.

This paper is an investigation of the pathological phenomena which can occur in commutative rings in the absence of the axiom of choice.

#12 L. Roberts, 1973. An example of a Hilbert ring with maximal ideals of different heights, Proc. Amer. Math. Soc. 37, 425–426.

W. Heinzer, 1974. Hilbert integral domains with maximal ideals of pre-assigned height, J. Alg. 29, 229–231.

The simplest examples of Hilbert rings are all rings of polynomials over fields, and the next simplest are the affine rings, which are homomorphic images of such polynomial rings. These rings, especially if they are domains, tend to have the property that all the maximal ideals are of the same height. Thus it is interesting to discover Hilbert domains having maximal ideals of different heights.

The first of these papers exhibits a Hilbert domain which is 2-dimensional, but still contains an uncountable number of maximal ideals of height 1. The ring involved is a suitable localization of $\mathbf{C}[X, Y]$.

Heinzer's paper presents a procedure for the construction of a Hilbert domain having maximal ideals of precisely the heights $r_1 < r_2 < \ldots < r_n$, where the r_i are any n given positive integers.

It should be recalled that in general a localization of a Hilbert ring is not Hilbert.

Also see Example 174 and [Gilmer 1971a].

#13 H. Butts and R. Yeagy, 1976. Finite bases for integral closures, J. Reine. Angew. Math. 282, 114–125.

This paper gives examples of (a) a non-Noetherian almost Dedekind domain R with quotient field K which is such that the integral closure of R in any finite separable algebraic extension L of K is a finitely generated R-module; and (b) a non-Noetherian almost Dedekind domain R with quotient field K such that there exists a finite separable algebraic extension L of K for which the integral closure of R in L is not a finitely generated R-module.

These examples should be considered in light of the behavior of Dedekind domains in such circumstances; see Chapter 3, Section D and Example 77.

#14 P. Cohn, 1968. Bézout rings and their subrings, Proc. Camb. Phil. Soc. 64, 251–264.

An integrally closed domain R is said to be a *Schreier* domain if any two factorizations of any given element have a common refinement (in the sense that 30 = (6)(5) and 30 = (3)(10) can be refined 30 = (2)(3)(5)). Clearly, any factorial domain is Schreier.

This paper deals with Schreier domains and Bézout domains. Among other things, it gives an example of a Schreier domain which is not pseudo-Bézout and an example of a domain which is not Schreier.

#15 J.Eagon, 1969. Examples of Cohen–Macaulay rings which are not Gorenstein, Math. Z. 109, 109–111.

The title is self-explanatory; the rings involved are localizations of appropriate affine rings.

#16 D. Ferrand and M. Reynaud, 1970. Fibres formelles d'un anneau local Noetherian, Ann. Sci. Ecole Norm. Sup. 3, 295–311.

Suppose R is a local domain with quotient field K. Let T be the completion of R. Then T may not be a domain (see Examples 103 through 109). For some time, it was unknown whether in T the ideal (0) might have embedded primes.

This paper gives an example of a 2-dimensional local domain R, such that in the completion T of R, (0) has an embedded prime (obviously the maximal ideal). There is also an example of a 1-dimensional local domain R with quotient field K, such that if T is the completion of R, then $T \otimes_R K$ is not Gorenstein.

#17 R. Gilmer and W. Heinzer, 1967. Intersections of quotient rings of an integral domain, J. Math. Kyoto Univ. 7, 133–150.

This paper deals with the QQR property and the concept of a unique minimal overring. (T is the *unique minimal overring* of R if T is a proper overring of R which is contained in all other proper overrings of R.) In this paper one can find an example of a QQR-domain which is not Prüfer and an example of a quasi-local QQR-domain having a unique minimal overring which is not a valuation domain. (This is of interest inasmuch as the integral closure of a

QQR-domain is Prüfer and, for many domains, the unique minimal overring, if it exists at all, turns out to be the integral closure.)

Also see [Gilmer 1973].

#18 M. Hochster, 1973. Contracted ideals from integral extensions of reg-
          ular rings, Nagoya Math. J. 51, 15–43.

A domain R is said to be *ideally integrally closed* (I.I.C.) if $IS \cap R = I$ for each of its integral extensions S and each of its ideals I. An I.I.C. domain is integrally closed, but the converse is false.

This paper examines the situation in which S is an integral extension of R, and asks when every ideal I of R is *contracted* from S (i.e., $IS \cap R = I$). One result is that the example constructed in [Bertin 1967] of a 4-dimensional, factorial, Noetherian domain which is not Cohen–Macaulay (see Example 129) is not I.I.C.

One related fact is that if S is a faithfully flat extension of R (e.g., a flat integral extension), then every ideal of R is contracted from S.

#19 G. Klatt and L. Levy, 1969. Pre-self-injective rings, Trans. Amer. Math.
          Soc. 137, 407–419.

A ring is said to be *self-injective* if it is injective as a module over itself. A ring is said to be *pre-self-injective* if all of its proper homomorphic images are self-injective.

This paper characterizes rings which are self-injective or pre-self-injective, giving examples of most of the major possibilities, some of which are rather surprising. For instance, there are self-injective rings of dimension greater than 0.

#20 T. Nakayama, 1942 and 1946. On Krull's conjecture concerning complete-
          ly integrally closed integrity domains, Part I. Proc. Imper. Acad.
          Tokyo 18 (1942), 185–187, Part II, *ibid.*, 18 (1942), 233–236,
          Part III, *ibid.*, 22 (1946), 249–250.

This paper gives an example of a completely integrally closed, infinite-dimensional, Bézout domain, none of whose overrings is a 1-dimensional valuation domain.

A similar example was later given by Ohm. See [Gilmer 1972a, pp. 236–238].

#21 B. Osofsky, 1969. A commutative local ring with finite global dimension and zero-divisors, Trans. Amer. Math. Soc. 141, 377–385.

This paper gives an example of a quasi-local ring (with 1) which is of finite global dimension and is not a domain. Of course, the ring is not Noetherian, since any local ring of finite global dimension is necessarily regular (and therefore a domain).

#22 J. Sally, 1972. A note on integral closure, Proc. Amer. Math. Soc. 36, 93–96.

This paper is concerned with the following question: If R is an integrally closed domain and x, y are nonzero elements of R, when is $R[x/y]$ integrally closed? Examples illustrate some of the possibilities.

#23 J. Sally, 1972. Regular overrings of regular local rings, Trans. Amer. Math. Soc. 171, 291–300.

Suppose R is a regular local ring and T is an overring of R, say for instance $T = R[x/y]$, where x, y are nonzero elements of R. It is reasonable to ask whether such a ring T is integrally closed or regular. This paper contains various examples of regular local rings and overrings which either are or are not integrally closed or regular.

#24 P. Sheldon, 1971. How changing $D[[X]]$ changes its quotient field, Trans. Amer. Math. Soc. 159, 223–244.

Among other matters, this paper considers the question of whether, for a domain R, the quotient field of $R[[X]]$ is determined entirely by the complete integral closure of R. Among other examples in the paper is an example of two distinct valuation domains V and W, having the same quotient field K, such that K is the complete integral closure of either one, but also that $V[[X]]$ and $W[[X]]$ have unequal quotient fields.

(There is more use made, in this discussion, of almost integral extensions than anywhere else I know.)

Index of Notations

Ann(x)	annihilator of an element x
Ass(I)	set of associated primes of an ideal I
Cl(R)	divisor class group of a completely integrally closed domain R
dim(R)	Krull dimension of a ring R
div(I)	divisor of a fractional ideal I
$D(R)$	monoid of all divisorial ideals of a domain R
g.c.d.(x, y)	greatest common divisor of two elements x and y
GD	going-down
Gd(R)	global dimension of a ring R
$G(R)$	group of divisibility of a domain R
Gr(I, A)	grade of an ideal I on a module A
Gr(R)	grade of a local ring R = Gr(M, R), where M is the maximal ideal of R
GU	going-up
ht(P)	height of a prime ideal P
I^{-1}	inverse of a fractional ideal I
id(M)	injective dimension of a module M
INC	incomparability
$I(R)$	group of all invertible fractional ideals of a domain R
$K((X))$	field of Laurent series in X over a field K
l.c.m.(x, y)	least common multiple of two elements x and y
LO	lying-over
pd(M)	projective dimension of a module M
Prin(R)	group of all principal fractional ideals of a domain R
$\overline{R}$	completion of a topological ring R
R^*	group of units of a ring R
R'	complete integral closure of a domain R
$R^{(n)}$	$(R^{(n-1)})'$
rad(I)	radical of an ideal I
R_g	subring of a ring R left fixed by an automorphism g
R_G	subring of a ring R left fixed by a group G of automorphisms
R_P	localization of a ring R at a prime ideal P
R_S	localization of a ring R at a multiplicatively closed set S
$S_1 \otimes_R S_2$	tensor product of two rings over a common subring R

tr. deg.(L/K)	transcendance degree of a field L over a subfield K
VNR	von Neumann regular
(x)	principal ideal generated by x

Index of Definitions